The Open University

MU120
Open Mathematics

GW00640933

Unit 8

Symbols

MU120 course units were produced by the following team:

Gaynor Arrowsmith (Course Manager)
Mike Crampin (Author)
Margaret Crowe (Course Manager)
Fergus Daly (Academic Editor)
Judith Daniels (Reader)
Chris Dillon (Author)
Judy Ekins (Chair and Author)
John Fauvel (Academic Editor)
Barrie Galpin (Author and Academic Editor)
Alan Graham (Author and Academic Editor)
Linda Hodgkinson (Author)
Gillian Iossif (Author)
Joyce Johnson (Reader)
Eric Love (Academic Editor)
Kevin McConway (Author)
David Pimm (Author and Academic Editor)
Karen Rex (Author)

Other contributions to the text were made by a number of Open University staff and students and others acting as consultants, developmental testers, critical readers and writers of draft material. The course team are extremely grateful for their time and effort.

The course units were put into production by the following:

Course Materials Production Unit (Faculty of Mathematics and Computing)

Martin Brazier (Graphic Designer)
Hannah Brunt (Graphic Designer)
Alison Cadle (TEXOpS Manager)
Jenny Chalmers (Publishing Editor)
Sue Dobson (Graphic Artist)
Roger Lowry (Publishing Editor)

Diane Mole (Graphic Designer)
Kate Richenburg (Publishing Editor)
John A.Taylor (Graphic Artist)
Howie Twiner (Graphic Artist)
Nazlin Vohra (Graphic Designer)
Steve Rycroft (Publishing Editor)

This publication forms part of an Open University course. Details of this and other Open University courses can be obtained from the Student Registration and Enquiry Service, The Open University, PO Box 197, Milton Keynes MK7 6BJ, United Kingdom: tel. +44 (0)845 300 6090, email general-enquiries@open.ac.uk

Alternatively, you may visit the Open University website at http://www.open.ac.uk where you can learn more about the wide range of courses and packs offered at all levels by The Open University.

To purchase a selection of Open University course materials visit http://www.ouw.co.uk, or contact Open University Worldwide, Walton Hall, Milton Keynes MK7 6AA, United Kingdom, for a brochure: tel. +44 (0)1908 858793, fax +44 (0)1908 858787, email ouw-customer-services@open.ac.uk

The Open University, Walton Hall, Milton Keynes, MK7 6AA.

First published 1996. Second edition 2003. Third edition 2008.

Edited, designed and typeset by The Open University, using the Open University TEX System.

Printed and bound in the United Kingdom by The Charlesworth Group, Wakefield.

ISBN 978 0 7492 2866 8

3.1

Contents

Study guide

This unit builds upon the work of previous units that have employed symbols. For example, the formula for Naismith's rule in *Unit 6* and the formula for a straight line in *Unit 7* both involved symbols.

Mathematics is often referred to as a language. Like other languages, it has its own grammar which you will need to learn. You may wish to listen to the preparatory audio band, 'Communicating mathematically', whilst studying this unit and/or refer to Module 1 of *Preparatory Resource Book A* for information about the role of brackets and other symbols in mathematics.

Section 1 of this unit has an associated audio band on number games; this introduces symbols and covers the use of brackets with numbers and with symbols. Section 2 considers the language of mathematical symbols. Sections 3 and 5 involve using your calculator as well as Chapter 8 of the *Calculator Book*. There is also an audio band for Section 5. Section 4 is concerned with manipulating expressions and equations on paper. There is no video for this unit.

The time that you need to spend on the unit is likely to depend very much upon your previous experience of using symbols, formulas and algebra. If you have little experience, then you should plan to spend more time than on previous units; you may also find it useful to get more practice by doing exercises from *Resource Book B*. On the other hand, if you are a confident user of algebra, you may be able to omit some of the activities.

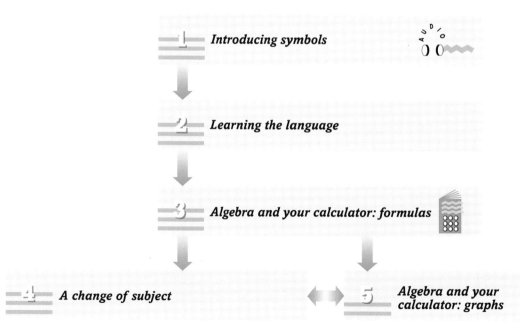

1   *Introducing symbols*

2   *Learning the language*

3   *Algebra and your calculator: formulas*

4   *A change of subject*

5   *Algebra and your calculator: graphs*

Summary of sections and other course components needed for *Unit 8*.

Introduction

It is quite common in ordinary life to represent things by symbols and other abbreviations. Examples include: 'Z' on a road sign warning of a double bend; map symbols such as the letter P to indicate parking, or a cross to indicate a church; chemical formulas such as H_2O for water. Symbols are concise: they save writing out a whole word or sentence, and consequently they often make it possible to see things more clearly.

The use of symbols constitutes a powerful mathematical technique. Some mathematical symbols will be very familiar to you, like $+$, $-$, $\times$, $\div$, $=$ and the decimal point. Also, you have already used the symbols $(,)$, $\sqrt{\ }$, x^2 and x^{-1} on your calculator.

Formulas or other mathematical expressions often employ letters as symbols in order to represent numerical quantities that can take different values. You have met the use of letters in the formula for the weighted mean and in the formula for a straight line. Frequently, the same formula can be written in several different ways, each suitable for a different purpose. This unit looks at methods of *manipulating* formulas and expressions into different but equivalent versions. You need to be able to perform such manipulations so that you can choose the most convenient version of a formula or expression for the purpose in hand. This area of mathematics is called *algebra*.

Symbolic and graphical representations are often linked, as discussed in *Unit 7* where the equation of a straight line was linked to its graph. It is convenient to be able to move easily from one representation to the other, and this unit will give you experience in that skill.

1 Introducing symbols

Aims This section aims to illustrate the usefulness of symbols and formulas and to introduce some of the rules for manipulating them. ◇

What do you think of when you hear the words 'symbols' and 'algebra'? Perhaps you immediately envisage complex strings of letters that mean very little to you. Or perhaps you have met symbols before in a mathematical or scientific context and have not been particularly bothered by them.

Activity 1 *Symbols and me*

Take a few minutes to think about your experiences of using symbols in mathematics up to now. Where have you used symbols in this course, and elsewhere?

Now, think about algebra. Jot down notes concerning your experiences of using symbols and algebra and your feelings about these topics, perhaps on the activity sheet provided.

You may find it helpful to add to this activity sheet as you work through the unit and to record the progress that you make. Note down ideas or techniques that you feel confident about, and others that you find difficult. Try to say why certain things are hard to understand and record how you sorted them out (if you managed to!): you may have asked another student, found out during a tutorial, or worked through the difficulty by yourself.

At the end of the unit you should be able to appreciate how far you have progressed in using symbols and algebra, by drawing on the evidence recorded on your activity sheet.

1.1 Number games

Listen to band 2 of CDA5509 (Tracks 2–14) while working through the following frames.

Frame 1

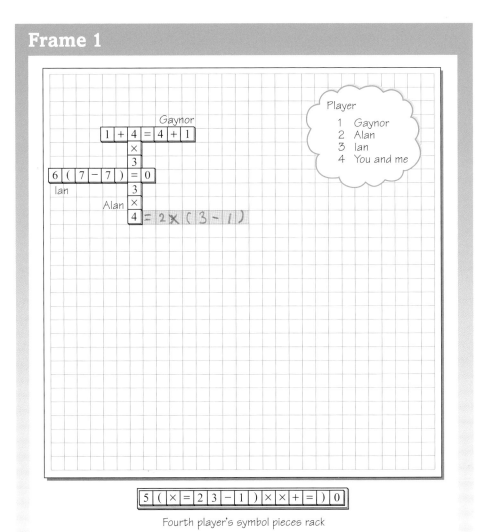

Fourth player's symbol pieces rack

Frame 2

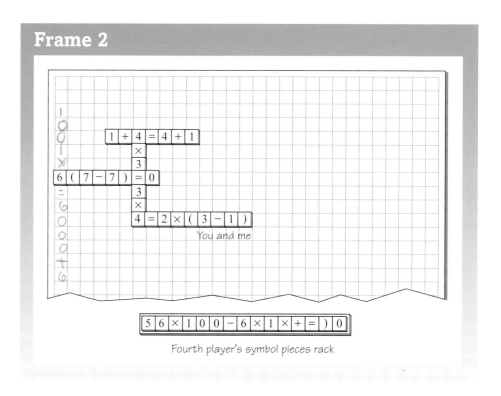

You and me

Fourth player's symbol pieces rack

Frame 3

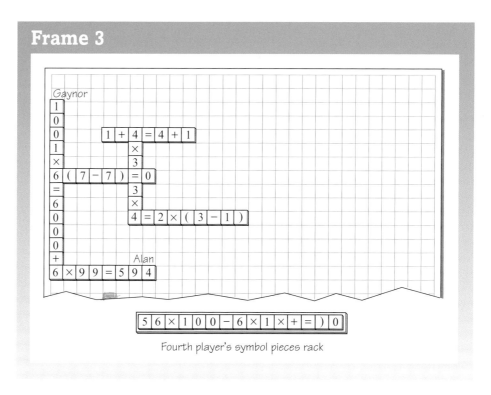

Fourth player's symbol pieces rack

Frame 4

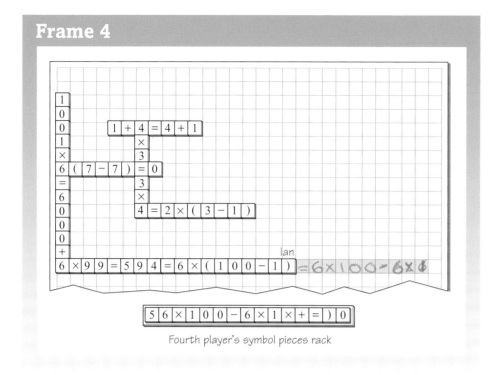

Fourth player's symbol pieces rack

Frame 5

Distributive property

$6 \times (100 - 1) = 6 \times 100 - 6 \times 1$

$(1000 + 1) \times 6 = 1000 \times 6 + 1 \times 6 = 6000 + 6$

Frame 6

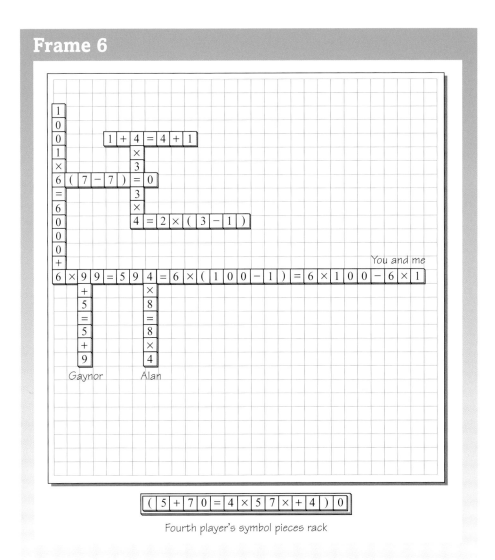

Fourth player's symbol pieces rack

Frame 7

Commutative property

Addition: $9 + 5 = 5 + 9$

Multiplication: $4 \times 8 = 8 \times 4$

Frame 8

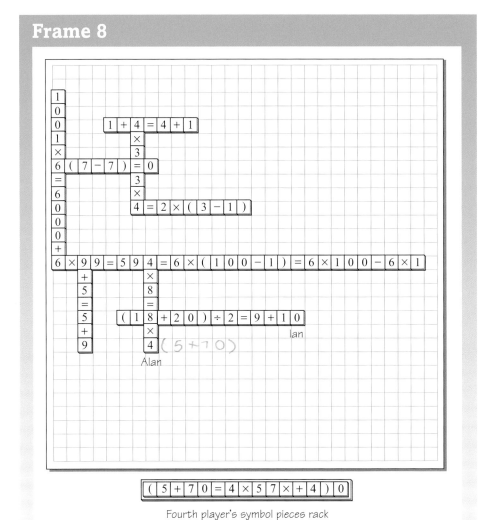

Fourth player's symbol pieces rack

$4 \times 5 +$
4×70

Frame 9

Distributive property

$(18 + 20) \div 2 = 18 \div 2 + 20 \div 2$

$\qquad\qquad\quad = \quad 9 + 10$

Frame 10

Distributive property

$$4 (5 + 70) = 4 \times 5 + 4 \times 70$$

or

$$4 (70 + 5) = 4 \times 70 + 4 \times 5$$

Frame 11

3692

$2 \times 3692 = 7384$

$+ 5 \qquad 7389$

14778

14772

3693

Think of a number

Double it

Add 5

Double the result

Subtract 6

Divide by 4

Take away the number you first thought of

5

$5 \times 2 = 10$

Frame 12

Think of a number		7
Double it	2×7	14
Add 5	$14 + 5$	19
Double the result	2×19	38
Subtract 6	$38 - 6$	32
Divide by 4	$32 \div 4$	8
Take away the number you first thought of	$8 - 7$	1

Frame 13

Think of a number	
Add 4	
Take away the number you first thought of	

Frame 14

Think of a number	7
Add 4	$7 + 4$
Take away the number you first thought of	$7 + 4 - 7$
And your answer is	4

Frame 15

Think of a number	7
Double it	2×7
Add 5	$2 \times 7 + 5$
Double the result	$2 \times (2 \times 7 + 5)$ or $2 \times 2 \times 7 + 2 \times 5$ or $4 \times 7 + 10$
Subtract 6	$4 \times 7 + 10 - 6$ or $4 \times 7 + 4$
Divide by 4	$(4 \times 7 + 4) \div 4$ or $4 \times 7 \div 4 + 4 \div 4$ or $7 + 1$
Take away the number you first thought of	$7 + 1 - 7$ or 1

Frame 16

Steps using the distributive property:

- $2 \times (2 \times 7 + 5)$

  The 2 multiplies *both* terms in the bracket to give

  $2 \times 2 \times 7 \quad + \quad 2 \times 5$.

- $(4 \times 7 + 4) \div 4$

  The 4 divides *both* terms in the bracket to give

  $4 \times 7 \div 4 \quad + \quad 4 \div 4$,

  which is

  $7 + 1$.

Frame 17

Think of a number	Number
Double it	$2 \times$ Number
Add 5	$2 \times$ Number $+ 5$
Double the result	$4 \times$ Number $+ 10$
Subtract 6	$4 \times$ Number $+ 4$
Divide by 4	Number $+ 1$
Take away the number you first thought of	Number $+ 1 -$ Number $= 1$

Frame 18

Think of a number	N
Double it	$2 \times N$
Add 5	$2 \times N + 5$
Double the result	$4 \times N + 10$
Subtract 6	$4 \times N + 4$
Divide by 4	$N + 1$
Take away the number you first thought of	$N + 1 - N = 1$

And your answer is, indeed, 1 whatever the value of N.

Frame 19

Compare

$\quad 2 \times (2 \times N + 5) = 2 \times 2 \times N \quad + \quad 2 \times 5 = 4 \times N \quad + \quad 10$

and

$\quad 2 \times (2 \times 7 + 5) = 2 \times 2 \times 7 \quad + \quad 2 \times 5 = 4 \times 7 \quad + \quad 10.$

Compare

$\quad (4 \times N + 4) \div 4 = 4 \times N \div 4 \quad + \quad 4 \div 4 = N \quad + \quad 1$

and

$\quad (4 \times 7 + 4) \div 4 = 4 \times 7 \div 4 \quad + \quad 4 \div 4 = 7 + 1.$

Activity 2 *Summarizing ideas*

The audio introduced some ideas that may be new to you. Write notes about these ideas, perhaps on your Handbook sheet. Among other things, you might want to cover commutative and distributive properties, expanding brackets, and using letters to stand for numbers in 'think of a number' tricks.

1.2 Formulas and functions

Formulas are often presented in either word or symbolic forms. For example, each line in the 'think of a number' trick gives rise to a *word* formula for generating the next number in the trick. Alternatively, each line can give rise to a *symbol* formula; thus the expressions on the right of Frame 18 are symbol formulas for generating these numbers and are written in terms of the number you first thought of, which is denoted by the symbol N. These two versions of a formula each have their own strengths and weaknesses. When given in words, formulas can be easier to remember, but when in symbols, they are easier to write and can more readily utilize the power of algebra.

Formulas are used in a very wide variety of contexts. You have already met quite a few formulas in this course, some in words and others in symbols. They include: formulas for the mean, weighted mean and standard deviation in Block A; and the formulas for Pythagoras' theorem and Naismith's rule in Block B.

An important aspect of a formula is that it shows a relationship between different quantities. For instance, in *Unit 6*, the application of Pythagoras' theorem led to the formula

See Section 4.2, *Unit 6*.

$$(\text{distance along road})^2 = (\text{horizontal distance})^2 + (\text{vertical distance})^2,$$

which shows the relationship between three distances.

The formula for Naismith's rule was given in *Unit 6* as

See Section 3.5, *Unit 6*.

$$\text{time taken for a walk} = \frac{\text{distance in kilometres}}{5} + \frac{\text{total ascent in metres}}{600}.$$

This formula can also be expressed in symbols. In this form the relationship between the approximate time t (in hours) for a walk, the distance walked d (in kilometres) and the total ascent h (in metres) is given by

$$t = \frac{d}{5} + \frac{h}{600}.$$

The formula above is written in a particularly useful way. The quantity you want to find, t, is on the left-hand side; it is equal to an expression on the right-hand side, involving the quantities d and h which you need to know in order to find t. So you put in values for d and h, and get out a value for t. The general principle is you input data and get an output.

In this format, a formula is called a *mathematical function*. There are *inputs* (in this case, the distance d and the ascent h) and an *output* (the time t), and there is a *rule* that has to be applied to the *process* of calculating the output from the inputs. These are the hallmarks of a mathematical function and are illustrated in Figure 1. Note that the word 'function' has this particular meaning when it is used in a mathematical sense.

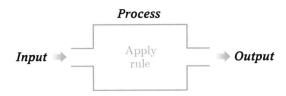

Figure 1   A mathematical function.

When the rule for the process of obtaining the output from the input in a function is written down, letters are often used to represent the input numbers and the output numbers, as in the formula given on the preceding page for Naismith's rule.

The application of Pythagoras' theorem shown opposite is not written in the form of a mathematical function because the left-hand side is the *square* of the output. However, it could be rewritten in a different form, involving a square root, so that it fulfils the criteria of a function:

$$\text{distance along road} = \sqrt{(\text{horizontal distance})^2 + (\text{vertical distance})^2}.$$

To summarize, a mathematical function is characterized by certain properties:

- It must have an *input*, a *process*, and an *output*.
- The output must be uniquely specified by the input.

A mathematical function is usually written in the form:

$$\text{output} = \text{expression involving input.}$$

You will find mathematical functions useful in many situations. In *Unit 7* you employed graphs for converting quantities from one set of units to another. Now, consider an alternative approach which uses mathematical functions.

Conversion formulas

The conversion graphs in *Unit 7* included changing from distances in miles to distances in kilometres, and from temperatures measured on the Fahrenheit scale to temperatures on the Celsius scale. Each of these conversion graphs has a corresponding formula which can be used instead of the graph.

Suppose that you need a formula for converting distances given in miles to kilometres, perhaps for a French friend who does not feel happy with the peculiarities of the British imperial system. To obtain the formula, you can use the fact that 1 mile is the same as 1.61 kilometres (to 2 decimal places). So to find the corresponding distance in kilometres from the number of miles, you multiply by 1.61. This is rather like an instruction in a 'think of a number' trick. The next thing to do is to choose a letter to stand for the *number* of miles: M seems an obvious choice. Then, the number of kilometres corresponding to M miles is $1.61 \times M$. It can also be useful to have a symbol for the number of kilometres: call it K. Therefore the formula for converting from miles to kilometres can be written as

$$K = 1.61 \times M.$$

Note that M stands for the *number of* miles *not* for miles, and K stands for the *number of* km *not* for km.

In words, this formula is

number of kilometres $= 1.61 \times$ number of miles.

Notice that K is a function of M. As Figure 2 indicates, the input is M (the number of miles), the output is K (the number of kilometres), and the rule is 'multiply by 1.61'. Thus $K = 1.61 \times M$.

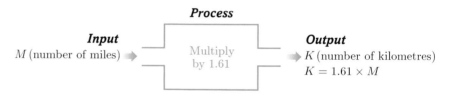

Figure 2   Representation of the function $K = 1.61 \times M$.

To find the equivalent in kilometres of a distance given in miles, you just replace M in the formula by the number of miles. For example, to convert 5 miles into kilometres, substitute 5 for M:

$$K = 1.61 \times 5 = 8.05.$$

Hence 5 miles is about 8 km.

Similarly, $M = 27$ gives $K = 1.61 \times 27 = 43.47$. So 27 miles is nearly 43.5 km.

In the UK, the long period of transition from the imperial system to the metric system has meant a lot of changing from one system of units to another, and therefore a number of conversion formulas are useful.

The UK started to go metric in the 1960s.

Example 1   *Pounds to kilos*

One pound (lb) is 0.45 kilogram (kg). Find a formula for converting from pounds to kilograms.

Solution

If the mass of something is P lb or J kg, then the formula needs to relate the numbers P and J.

As 1 lb is the same mass as 0.45 kg, it follows that P lb is $0.45 \times P$ kg.

However, P lb is the same as J kg. So

$$J = 0.45 \times P.$$

This is the required formula.

It is better not to use K again, to avoid confusing kilograms with kilometres. (From an algebraic point of view, it is a nuisance that a number of metric measure words begin 'kilo'.)

Activity 3   *Gallons and litres*

(a) One (imperial) gallon is 4.55 litres. Find a formula for converting from gallons to litres.

(b) Milk and beer are often sold in pints in the UK. One pint is an $\frac{1}{8}$ of a gallon. What is this in litres?

Example 2   *Kilometres to miles*

Some British people are unused to measuring distances in kilometres. Imagine you have some such friends, who are planning to travel in mainland Europe, using a map on which all the distances are given in kilometres. How might you help them to convert these distances into miles?

(a) Find a formula like the one for converting miles to kilometres, but the other way round.

(b) Use the formula to convert 40 km to the corresponding number of miles. Check whether the formula is correct by converting your answer back to kilometres, using $K = 1.61 \times M$.

Solution

(a) As 1 mile is the same as 1.61 km, it follows that 1 km is $1 \div 1.61$ miles.

Therefore, K km is $K \div 1.61$ miles.

But K km is the same distance as M miles, so

$$M = K \div 1.61.$$

(b) Substituting 40 for K in this formula gives

$$M = 40 \div 1.61 = 24.84 \text{ (to 2 decimal places).}$$

Hence 40 km is about 24.84 miles.

To convert 24.84 miles back to kilometres, substitute 24.84 for M in $K = 1.61 \times M$:

$$K = 1.61 \times 24.84 = 40.00 \text{ (to 2 decimal places).}$$

Hence 24.84 miles is about 40 km, and this shows that the formula works.

Activity 4   *Litres to gallons*

When thinking about petrol consumption, some British people still prefer to use gallons rather than litres. Find a formula that you could enter into your calculator to convert from litres to gallons.

Use the values given in Activity 3.

Use the formula to check your answer to Activity 3(b).

Example 2 and Activity 4 show that sometimes a formula can be reversed. This gives what is called an *inverse mathematical function*. In such cases, the output from the function becomes the input for the inverse function, and the output of the inverse function is the input for the original function (see Figure 3 opposite).

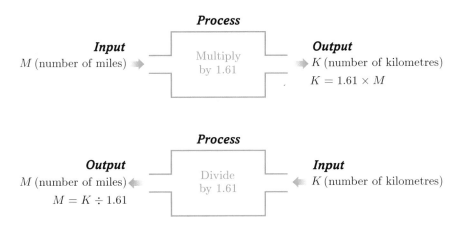

Figure 3   A function and its inverse function.

However, it is not always possible to reverse a formula and obtain an input that corresponds to a given output. For instance, it is not possible to reverse Naismith's rule and find both the distance covered and the height climbed when all you know is the estimated time for a walk. This is because the same output can be obtained from many different inputs: a short steep climb may have the same estimated time as a longer flat walk.

When formulas can be reversed there are techniques for producing the inverse mathematical function. These will be dealt with in detail in Section 4.

Activity 5   *How tall was Goliath?*

The unit of measurement of length used by many ancient peoples, including Egyptians, Babylonians, Greeks, Romans and Hebrews, was the *cubit*. It was based on the length of the arm from the elbow to the extended fingertips, and is usually taken to be about 0.46 m. Therefore the formula relating the number of cubits, C, in a given length to the number of metres, M, in that length is $M = 0.46 \times C$.

According to the Bible (1611 translation): ... 'there went out a champion out of the camp of the Philistines, named Goliath, of Gath, whose height was six cubits and a span'. (I Samuel 17:4)

Assume that a span is half a cubit (the span of a hand) and work out roughly how tall Goliath was in metres.

Now explain how you did this as if to a friend who does not know any algebra. Note down the advantages of using a formula instead of words.

A more complex conversion formula than those considered so far is the one needed to convert a temperature from degrees Fahrenheit to degrees Celsius. This is because, in addition to the units being different sizes, the zeros on the two temperature scales do not coincide.

In *Unit 7*, two fixed points were used to draw a graph for this conversion: these points were the temperature at which water freezes (32 °F, or 0 °C) and the temperature at which water boils (212 °F, or 100 °C). The same data may be used to construct a formula for converting temperatures.

Suppose you want to convert a temperature of, say, 86 °F into degrees Celsius. In Figure 4, you can see that 86 °F is 54 Fahrenheit degrees above the Fahrenheit freezing point of 32 °F (54 is found by subtracting 32 from 86).

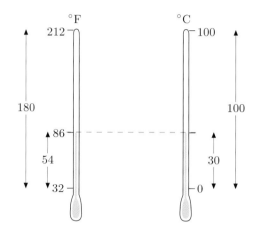

Figure 4   Fahrenheit and Celsius thermometers.

Each of these 54 Fahrenheit degrees needs to be converted into an equivalent number of Celsius degrees. To do this, notice that *between* the freezing and boiling points of water on the Fahrenheit scale, there is a rise of 180 °F, which is equivalent to a rise of 100 °C on the Celsius scale. Thus:

Fahrenheit		Celsius
A rise of 180 °F	is equivalent to	a rise of 100 °C
A rise of 1.8 °F	is equivalent to	a rise of 1 °C
A rise of 1 °F	is equivalent to	a rise of $1 \div 1.8$ °C

So, to change each Fahrenheit degree to an equivalent number of Celsius degrees, you need to divide by 1.8. This means that 54 Fahrenheit degrees above the Fahrenheit freezing point are equivalent to $54 \div 1.8 = 30$ Celsius degrees above the Celsius freezing point. As the Celsius freezing point is 0 °C, the Celsius equivalent to a temperature of 86 °F is 30 °C.

This method will work for any temperature conversion from Fahrenheit to Celsius. The procedure is a two-step process:

● subtract 32 (to get the number of degrees above freezing);

● divide the result by 1.8.

This pair of instructions reads rather like part of a 'think of a number' sequence, with the Fahrenheit temperature taking the role of the number you first thought of. A similar procedure to that used in the audio frames sequence will then produce a formula that gives the equivalent on the

Celsius scale, $c\,°\mathrm{C}$, of a temperature, $f\,°\mathrm{F}$, on the Fahrenheit scale. In the first step, subtract 32 from f to obtain $f - 32$. In order to avoid any ambiguity, put this result in brackets, which gives $(f - 32)$. In the second step, divide the result of step 1, that is $(f - 32)$, by 1.8. So the final formula is

$$c = (f - 32) \div 1.8.$$

This is a mathematical function involving two steps, as shown in Figure 5. Here a Fahrenheit temperature is the input, and the equivalent Celsius temperature is the output.

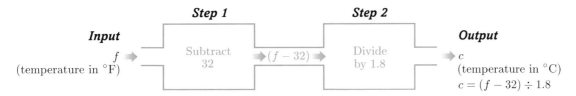

Figure 5   The two-step formula for converting between °F and °C.

Example 3   *Checking boiling points*

Check the Fahrenheit to Celsius conversion formula by seeing if it produces $100\,°\mathrm{C}$ as the temperature that corresponds to $212\,°\mathrm{F}$, the boiling point of water.

Solution

In order to check the conversion formula, substitute 212 for f. This gives

$$c = (212 - 32) \div 1.8,$$

so

$$c = 180 \div 1.8 = 100$$

as expected.

Like the other conversions, the Fahrenheit to Celsius conversion can be carried out in either direction, as the next activity demonstrates.

Activity 6   *A temperature conversion*

(a) Old medical books often quote the normal human body temperature as 98.4°F. Use the formula $c = (f - 32) \div 1.8$ to work out its Celsius equivalent.

(b) The steps in converting from the Fahrenheit to Celsius measurement of a temperature need to be reversed when doing the conversion the other way round: starting with the temperature on the Celsius scale, first multiply by 1.8, then add 32 to the result. This gives the formula

$$f = (c \times 1.8) + 32.$$

Use this formula to convert $50\,°C$ into Fahrenheit.

Check your answer by converting it back to Celsius, using

$$c = (f - 32) \div 1.8.$$

Activity 7   *Handbook activity on formulas and functions*

Make some notes about formulas and functions on your Handbook sheet, using examples if you find them helpful. Continue to add to these notes during the rest of this section, as and when appropriate.

Notice the different roles of the letters F and f, and C and c in the discussion of temperature conversion. Here F is an abbreviation of the name 'Fahrenheit' and labels the type of degrees in use; similarly, the letter C stands for Celsius. These are merely abbreviations for the units of measurement. On the other hand, f and c are symbols for *numbers*—the *number* of degrees Fahrenheit or Celsius, respectively. These symbols take on different values, depending upon the particular temperature to be converted. Sometimes such symbols are called *variables*, because their value varies.

When letters are used for variables, they are generally written in italics. When they represent units, they are not.

Notice also that lower-case letters have been used for f and c to avoid confusion with the units $°F$ and $°C$, which are in upper-case. For the same reason, in the conversions between miles and kilometres, where the units m and km are written in lower-case, the variables M and K are written in capital letters. In any context it is important not to use the same symbol for more than one thing as this can lead to confusion.

Although it is convenient to use letters that have some affinity with what they stand for, it is not necessary to do so. The formula for converting from miles to kilometres could just as well be written as $A = 1.61 \times B$ or $y = 1.61 \times x$, rather than as $K = 1.61 \times M$. Remember that it is the *form* of a formula or mathematical function (in this case, multiply one number by 1.61 to get the other) that distinguishes it, not the actual letters that appear in it.

When using symbols, you need to state clearly what each symbol means at the beginning of a new piece of work—and not change what it stands for as you go along!

You have a wide choice of letters from which to choose symbols. There are, in effect, 52 letters in the (Roman) alphabet (capital letters are usually considered as different from lower-case letters). You can also call on the Greek alphabet, although some letters are reserved for specific uses (for example, π is the ratio of the circumference of a circle to its diameter).

The Greek alphabet is given on POM Handbook Sheet 4.

Formulas for travel

Converting speeds from one set of units to another may involve converting both distances and times, because speed is a measure of distance travelled in a given time. To convert distances, use the methods introduced earlier; thus, to change a speed given in kilometres per hour (km/h) into miles per hour, the distance measurement needs to be converted by using the formula $K = 1.61 \times M$. When the unit of time also has to be changed, you need to divide or multiply as appropriate; therefore, in changing a speed in km/h into km/min, you would need to *divide* by 60 (as there are 60 minutes in an hour); in going the other way, you would need to *multiply* by 60.

The units km/h are also written $\mathrm{km\,h^{-1}}$.

Activity 8 *Eurostar*

(a) After the Eurostar train leaves the Channel Tunnel *en route* for Brussels or Paris, it passes through Calais-Frethun station; shortly after that, the conductor announces over the intercom: 'Ladies and gentlemen, the train is now travelling at its maximum speed of 300 kilometres per hour'. Use a suitable formula to convert this into miles per hour.

(b) In many scientific uses, speeds are measured in metres per second (m/s). Devise a formula for converting from km/h to m/s.

The units m/s are also written $\mathrm{m\,s^{-1}}$.

(c) What is a speed of 300 km/h in m/s?

In *Unit 7*, you met three word formulas relating average speed to distance travelled and travel time (time from the start of the journey). They were:

distance travelled is equal to average speed multiplied by travel time,

average speed is equal to distance travelled divided by travel time

and

travel time is equal to distance travelled divided by average speed.

These formulas can also be expressed in symbols. With the symbol d for the numerical distance travelled (measured from the starting point, in appropriate units), t for travel time (measured from the starting time, in appropriate units), and s for average speed (measured in consistent units), the formulas can be written as

Note that d, t and s need to be measured in consistent units; thus, if d is in km and t in hours, then s is in km/h.

$$d = s \times t, \quad s = \frac{d}{t} \quad \text{and} \quad t = \frac{d}{s}.$$

These formulas express the fundamental relationships between distance travelled, average speed and time taken, and many problems concerned with motion and travel can be solved with their aid. To use them for a calculation in a given case, you need to know the values of two of these quantities, as in the example overleaf.

Example 4   *Coach travel*

Suppose you were planning a coach journey across Italy, mostly along autostrada (motorways), and that you expected to average 95 km/h on a particular morning.

(a) What is the formula that gives the distance travelled by the coach when you input the time for a journey?

(b) Use the formula from part (a) to find how far you would expect to travel in $1\frac{1}{2}$ hours.

(c) What is the formula that gives the time taken by the coach when you input the distance travelled? Use it to find how long you would expect the coach to take to cover 264 km.

Solution

(a) The average speed of the coach is 95 km/h, hence $s = 95$. The general formula for the distance travelled d km in a time t hours is $d = s \times t$. In this case,

$$d = 95 \times t.$$

(b) Putting $t = 1\frac{1}{2}$ in the formula obtained in part (a) gives $d = 95 \times 1.5 = 142.5$. So you would expect to travel just over 140 km in $1\frac{1}{2}$ hours.

(c) The formula for the time t hours taken by the coach to travel a distance d km is

$$t = \frac{d}{95}.$$

When $d = 264$, then $t = 264 \div 95 = 2.78$ (to 2 decimal places). So to travel 264 km would take about $2\frac{3}{4}$ hours.

Activity 9   *Travelling fast*

(a) For an average speed of s km/h, find a formula for the time taken (in seconds) to travel 1 km.

(b) Use the formula from part (a) to find out how long (in seconds) each of the following takes:

(i) A Eurostar train covering a kilometre at its maximum speed of 300 km/h.

(ii) The Earth travelling a kilometre in its orbit through space. (The Earth's orbital speed is about 107 000 km/h.)

(iii) Light travelling a kilometre. (The speed of light is about 1 100 000 000 km/h.)

The formulas connecting distance, speed and time need to be amended when distances are not measured directly from the starting point of the journey, or time is not measured from the starting time. This is illustrated in the case described below.

Two colleagues, Alastair, who works in Cardiff, and Bethan, who works in Edinburgh, regularly make long car journeys: Alastair goes to Edinburgh, and Bethan to Cardiff. On occasions, when they are both travelling on the same day, they use the same route, and like to meet on the way. On the next such occasion, Alastair plans to leave Cardiff at 8.30 a.m. and drive at an average speed of 50 mph. Bethan plans to leave Edinburgh at 10 a.m. and drive at an average speed of 60 mph. The distance between Cardiff and Edinburgh is almost exactly 400 miles.

Later (in Section 4.3) you will be asked to estimate the time when they are likely to meet. The task for now is to derive formulas for where each of them will be at a given time. Since, in Activity 10, you will have to compare their distances and times, it is best to measure their distances from the same point and their times from the same time. Take Cardiff as the reference point for distances, and 10 a.m. as the reference point for times.

You could use either place and either starting time as the reference points.

First, find a formula for Alastair's distance from Cardiff, say a miles, at a time t hours after 10 a.m. At 10 a.m., he will already have been travelling for $1\frac{1}{2}$ hours, at an average speed of 50 mph; he will therefore be about 75 miles from Cardiff. With each further hour that passes, he will get about 50 miles further away from Cardiff. So in t hours he will be a further $50 \times t$ miles from Cardiff. Hence his approximate distance from Cardiff at t hours after 10 a.m. is given by the formula

$$a = 75 + (50 \times t).$$

The expression 't hours after 10 a.m.' is potentially a bit misleading: it sounds as though the formula is restricted to whole numbers of hours, but this is not so. To find how far Alastair will be from Cardiff at, say, 1.45 p.m., you take t to be $3\frac{3}{4}$, or 3.75.

When using the formula for times that are not whole numbers of hours, you have to remember to convert minutes to hours.

What about Bethan? Since she will be travelling at an average speed of 60 mph, the distance she will have gone in t hours will be about $60 \times t$ miles. However, she was 400 miles from Cardiff initially, and getting closer all the time. Her distance from Cardiff, say b miles, at t hours after 10 a.m. is given by

$$b = 400 - (60 \times t).$$

Activity 10   *How far from Edinburgh?*

Who would be closer to Edinburgh at 2.15 p.m., Alastair or Bethan?

Activity 11   *Average speeds*

Suppose you set off in a car for a 300-mile journey as a passenger. After a little while you notice that the time is 10.00 (on the 24-hour clock), and the trip meter (which the driver can never remember to reset) registers 100 miles. You begin to wonder how you could estimate the average speed later on in the journey from subsequent readings of the mileage on the trip meter. You decide to amuse yourself by devising a formula to work out the average speed (since 10.00) when the time is t hours (on the 24-hour clock) and the trip meter registers d miles.

(a) What is the formula for the average speed, s mph, in terms of t and d?

(b) Suppose that at a quarter past eleven (in the morning) the trip meter reads 180 miles. What has been your average speed since 10.00?

Other formulas

Formulas are useful in a variety of other situations. The following activities will help you to consolidate what you have just learned about using and devising formulas to describe different situations.

Activity 12   *Garden design*

These formulas are given on POM Handbook Sheet 3.

Suppose that you are planning a garden, and your plans include a lawn consisting of a rectangle 4.2 m by 3.6 m with two semi-circular areas of diameter 3.6 m added on to the shorter sides. Draw a plan of the lawn, and use the area formulas for a rectangle and a circle to estimate how much turf you will need to cover the area, giving your answer in square metres.

Activity 13   *Areas and volumes*

(a) The volume of a sphere can be expressed in words as follows: four thirds of π multiplied by the cube of the radius. Write this formula in symbols, taking the radius as r and the volume as V.

(b) A cylinder has radius r and length l. Its total surface area can be regarded as being made up of three parts: two circular ends and the curved surface. The curved surface can be cut vertically and unrolled to make a rectangle. This rectangle has its length equal to the length of the cylinder, and its width equal to the circumference of the circles at each end of the cylinder. Write down a formula for the total surface area of the cylinder in terms of r and l.

(c) The volume V of a cylinder is the area of its base multiplied by its length. Use the same symbols as in part (b) to express this word formula concisely in symbols.

Activity 14   *Party time*

(a) Niels is going to a fancy dress party and is planning to go as a red UK cylindrical post box. He will have to buy some red card to make a cylinder of height about 1 m and radius about 0.25 m. His post box will have a top but no base. What will be the area of the curved outer surface of his post box, and what will be the area of the top?

The red card comes in the following sizes: 1 m square, 1 m by 1.5 m, 1.5 m square, 1 m by 2 m, and 1.5 m by 2 m. Which should he buy?

(b) Sujatha has also been invited to the fancy dress party, and wants to go as a can of fizzy lemonade. She is planning to make a large version of a drinks can, with a height of 0.8 m. She has a piece of strong, silver-coloured card from which to make the curved part of her costume. Her piece of card is a rectangle 1 m by 1.1 m. What will be the radius of the can that she is able to make?

An interesting use of a formula is provided by Hooper's rule, which relates to the age of hedges. Some English hedges are very old and, due to self-seeded trees and shrubs, they contain several different species. In 1974, Dr Max Hooper obtained data on the numbers of different species of tree and shrub in 227 established English hedges whose ages he knew from written records. He fitted a formula to his data. This formula, now known as Hooper's rule, has been used successfully to estimate the ages of other English hedges.

To find the age of a hedge using Hooper's rule:

- mark the hedge off into consecutive 30-yard sections;
- count the number of different species of tree and shrub in each section;
- find the average (mean) value of the number of species;
- multiply this average value by 110, and add 30 to the result.

This gives the approximate age of the hedge in years.

Activity 15   *Hooper's rule as a formula*

(a) Express Hooper's rule as a formula, using A for the age of the hedge and N for the number of species, on average, in a 30-yard section of the hedge.

(b) A member of the course team recently carried out the procedure for eight sections of a hedge along the road from Aspley Guise to Husborne Crawley, two villages in Bedfordshire. The average number of species per 30-yard section was 3.5. Estimate the age of the hedge, according to Hooper's rule.

This section has introduced some of the basic ideas of algebra—most importantly, the notion of using a letter to stand for an unknown number. This idea was first employed to help you understand 'think of a number' tricks: the result of each step in the trick was represented by a formula involving N, the number first thought of. These tricks work because the final formula always reduces to a number that is independent of N. By using an algebraic formulation and then simplifying and rearranging it, you can see, more easily than by other methods, *why* the result will be the same irrespective of the number that is first thought of.

The same fundamental idea of representing a sequence of calculations by symbols underlies the use of formulas in many other situations: examples are formulas to represent relationships between measurements in different units, or between distances and times in travel problems.

Your work, both with 'think of a number' tricks and with formulas, has consisted of taking a sequence of operations described in words, and converting the sequence into a symbolic form with letters used to represent the variable quantities and with other symbols used to specify the arithmetic operations. This has involved a process of translation, from the English language into a symbolic language—the language of algebra.

Activity 16   *Pros and cons of algebra*

Draw up a list of the advantages and drawbacks of algebra, using the activity sheet provided. List as many advantages as you can.

Outcomes

After studying this section, you should be able to:

◇   represent the result of each step of a 'think of a number' trick by a formula involving a letter, such as N, to stand for the initial number, and hence explain why the trick works (or not) (Activity 2);

◇   understand the terms 'function', 'input' and 'output' (Activity 7);

◇   convert relationships (for example, between quantities such as equivalent measurements in different units) specified in words, into a symbolic form, making your own choice of letters to stand for the quantities involved, and use such formulas (Activities 3–6 and 8–14);

◇   recognize the benefits and drawbacks of the use of symbols in mathematics (Activities 5 and 16);

◇   consider your progress in using symbols, and record important concepts and techniques (Activities 1, 2 and 7).

2 *Learning the language*

Aims   The aim of this section is to enable you to say with greater
conviction: 'Mathematics—I speak it fluently'.   ◇

2.1 *The language of mathematics*

In Section 1, several word problems were translated into formulas written
in a concise symbolic form. You could think of this symbolic form as part
of the language of algebra. This language is used for expressing
mathematical ideas, generalizations and relationships. In order to study
mathematics successfully, you need to learn to use this language with
confidence.

A language includes words, as well as rules for the ways in which those
words can be put together to create meaningful phrases and sentences.
Change the arrangement of the words and you frequently change the
meaning. Hence, word order is important: 'The cat sat on the mat' means
something different from 'The mat sat on the cat'. Like all languages,
algebra has its own vocabulary and rules of grammar. This section is
devoted to discussing some of this vocabulary and grammar so that you
can develop greater fluency in reading and writing algebra.

In general, the 'words' of algebra are individual symbols. There are two
types of symbols: letters, which stand for numerical quantities, and signs,
which represent mathematical operations such as addition, multiplication
or raising to a power. The various symbols can be strung together to form
the equivalent of phrases and sentences by using the rules of algebraic
grammar. The rules of grammar for a language ensure that communication

is clear and meaningful: grammatical phrases have a better chance of being understood than ungrammatical ones. In algebra, grammar is even more important than in ordinary language: an ungrammatical algebraic phrase usually makes no sense at all.

Algebra is a concise language that is well adapted to making general statements, and to making them precisely and concisely. For example, by letting N stand for 'the number you first thought of', you can summarize *all* the possible results of carrying out the sequence of calculations in a 'think of a number' trick. Similarly, by letting f stand for an unspecified Fahrenheit temperature, you can write down a formula for converting *any* temperature from Fahrenheit to Celsius units. It is this type of concise generalization that makes algebra such a powerful tool.

The work of scientists like Newton and Einstein would have been severely limited without algebra. A famous example of the use of algebra is Einstein's equation $E = mc^2$, which is a concise way of saying that the energy E released in, say, a nuclear reaction equals the change in mass, m, multiplied by the square of the speed of light, c.

An important point in algebra is that the operations of arithmetic apply in just the same way to the letters as they do to numbers, because the letters, in fact, represent numbers. The operations include addition, subtraction, multiplication, division, and finding squares and square roots. In algebra, some of these operations are written slightly differently from in arithmetic, but their meanings are the same.

When learning any language, you need to practise using it at every available opportunity, and the same principle applies to learning algebra. Many activities in this section and in the *Resource Book* are designed to let you to practise your algebraic skills.

2.2 *Expressions*

A collection of symbols, such as $4 \times N + 10$, $1.61 \times M$ or $1.8 \times c + 32$, is called an *algebraic expression*, or simply an expression. The key feature of an algebraic expression is that it contains one or more symbols standing for unspecified numbers. The symbols representing unspecified numbers are often called *variables*.

An expression is the algebraic equivalent of an English phrase. It can be thought of as a way of representing a calculation process. Thus, the expression $4 \times N + 10$ represents the process 'take your number (whatever it is), multiply it by 4, and add 10 to the result'. If you then replace N in the expression $4 \times N + 10$ by, say, 0.5, you will get $4 \times 0.5 + 10$, which is 12.

Replacing a variable by a number in an expression is called *substitution*, and the number that replaces the variable is referred to as the *value of the variable*. So, in the above case, 0.5 was substituted for the variable N in the expression $4 \times N + 10$, and therefore the variable N is said to have the value 0.5.

You might like to add *substitute* and *value* to your Handbook sheet.

When the expression $4 \times N + 10$ is evaluated by substituting 0.5 for N, it gives an output of 12; in other words, the value of the expression is 12. Calculating the output by substituting values of the variables in an expression in this way is referred to as *evaluating the expression*. The output, or number you get as a result, is called the *value of the expression*. Sometimes the term 'numerical value' is used instead of 'value', to emphasize that the value is a number (or sometimes a number with units attached, as in 1.61 km).

For convenience, an equals sign is often used when giving a value to a variable and when presenting the subsequent result of evaluating an expression. An example is:

When $N = 0.5$, then $4 \times N + 10 = 12$.

Activity 17 *Evaluation*

(a) Complete the sentence: 'When $N = {}^-1$, then $4 \times N + 10 = \ldots$'.

(b) Complete the sentence: 'When $x = 6$, then $1 + \frac{1}{3} \times x = \ldots$'.

(c) Complete the sentence: 'When $y = 3$, then $12 \div y + 4 = \ldots$'.

Activity 18 *Values*

You could use your calculator for these questions if you wish. For instance, in (b), store the value of p in memory P (using ▶P) and then enter the formula.

(a) What is the value of $3 \times M + 5$ when $M = 3$?

(b) What is the value of $2 \times p - 2$ when p has the value

   (i) 3,   (ii) 7.165,   (iii) $^-1$?

(c) What is the value of $2 + b - 2 \times b$ when b is 5?

(d) When $a = 3$ and $b = 5$, what is the value of $a \times b - a - b + 1$?

Making sense in algebra

A collection of symbols like $4 \times N + 10$ constitutes an algebraic expression. But some collections of symbols make no algebraic sense: a case in point is $2 - \times M + \times N$. Here the grammar is wrong: in '$2 - \times M$', there is no indication as to what operation to carry out on 2, nor what to do to M; similarly, '$+ \times N$' is an incomplete instruction. A meaningless collection of symbols is not an algebraic expression, just as a meaningless collection of words would not be called an English phrase or sentence.

You will have to deal with quite complicated expressions as the course progresses; it is therefore important to learn to scan collections of symbols to check that they make sense. The question to ask yourself is this: when I substitute numerical values for the variables, will I be able to calculate a numerical result? If you try substituting $N = 0.5$ in $4 \times N+$, you soon see that you do not get a numerical value for the result! Each of the operations $\times$, $\div$, $+$ or $-$ must have a number after it to multiply by, divide by, add or subtract, though an expression may start with a negative sign ($^-$) when it signifies a negative number.

Some people like to read the expression aloud to see whether it makes sense or whether there are some words missing.

Activity 19 *Sense or nonsense*

Which (if any) of the following collections of symbols do not 'make sense'?

(a) $a + 2 -$ (b) $6 \times m + 35 - m$ (c) $+4 \times N - 10$

(d) $^-4 \times N + 15$ (e) $N \times \div 2$ (f) $(p + q) \div \div 2$

(g) $4 + 2 - \times 3$

Some algebraic conventions

In arithmetic and algebra, multiplication is indicated by the sign $\times$. However, it is not always necessary to use this sign to show that multiplication is required: for instance, $5(3+2)$ means $5 \times (3+2)$. It is normal practice in algebra to omit multiplication signs: thus you would write $4N$ for 4 multiplied by N, instead of $4 \times N$. This convention applies when *a number multiplies a symbol*, though not when a number multiplies another number: 24 continues to mean twenty-four, not 2 times 4, because the decimal system for representing numbers makes use of the relative positions of the digits. But note that an expression such as $2ab$ is a concise way of writing $2 \times a \times b$ and is not 'two hundred, a tens, and b units'.

Since $a \times b = b \times a$, it follows that $2 \times b \times a$ is also $2ab$.

When *a number multiplies an expression that begins with a number*, you should use a multiplication sign, as in $2 \times 4N$. Omitting the multiplication sign here will not do: $24N$ means twenty-four times N, not 2 times $4N$. So to avoid ambiguity, write $2 \times 4N$. To check this, store any number N in the memory of your calculator; then confirm that $2 \times 4N$, $4N \times 2$ and $8N$ all have the same value, but that $24N$ has a different value.

Multiplication is sometimes represented by a dot, as in $4 \cdot N$ meaning $4 \times N$, but this notation is not used in MU120.

Note that it is not wrong to use multiplication signs in algebra; it is just conventional to omit them in order to save writing, except when there could be ambiguity. If you find that putting in multiplication signs helps you to understand algebraic expressions, by all means do so.

Another algebraic convention concerns *the ordering of numbers and symbols*. You could write $N \times 4$ as $N4$ but it is usual to write the number first and the symbol second: thus $4N$ is preferable to $N4$. This avoids an ambiguity with other kinds of notation: for example, on your calculator, $L4$ means 'list 4'.

A further reason for putting the number first arises when using negative numbers. To appreciate this, think about the best way of writing, say, $^-4$ multiplied by N. The convention of putting the number first gives ^-4N. Whereas if you put the number second, it could easily be misread as $N - 4$, which means 'N minus four' and is quite different. Of course, you could write $N(^-4)$ to avoid the ambiguity, but if you adopt the 'number first' convention, the problem never arises.

You might like to make notes on these points on your Handbook sheet.

Many other symbols that will be familiar to you from arithmetic are used in algebra. The product of a number with itself, $p \times p$, is written as p^2 (it is not usually written as pp) and is called 'p squared'. Similarly, p^3, which is called 'p cubed', represents $p \times p \times p$, while $7R^2$ is $7 \times R \times R$. Expressions like p^2, p^3, and p^4 are referred to as *powers* of p. (Note that the variable itself is considered as the first power, so p can be thought of as p^1.)

Root signs are also used in algebra just as they are in arithmetic: for example, $\sqrt{N}$ is the (positive) square root of the number N (N is assumed to be positive); $\sqrt{5x^2}$ is the square root of $5x^2$; $\sqrt[3]{k}$ is the cube root of k; and $\sqrt[3]{p^2}$ is the cube root of the square of p.

Activity 20   Down with multiplication

Express the following without multiplication signs.

(a) $3 \times N$ (b) $4 \times \frac{1}{6} \times z$ (c) $^-2 \times 2N$ (d) $24 \times 2x$

(e) $^-5 \times 2N$ (f) $^-6 \times \frac{1}{3}x$ (g) $8P \times 7P$ (h) $5x \times 2x^2$

Activity 21   Doing exercises

How many of the eight questions in Activity 20 did you do? Did you do them all? Did you just look at them and think 'I know I can do these'? Did you do a few and skip the rest?

Did you find all the questions equally easy? Were some of a different kind?

Take a few moments to reflect on how you dealt with Activity 20. You do not need to write anything down, but read the comments below.

Activity 20 was a collection of questions on a particular theme: it constituted an exercise. The main purpose of an exercise is to enable you to practise a skill, and you will come across many such sets of exercises in this Unit and later in the course.

These exercises are often graded in difficulty, with easy questions at the start and harder ones at the end. As it is impossible for the people who wrote them to know how skilful you are at a certain type of question, only you can decide how much of the exercise you need to do. If the questions concern a skill that is unfamiliar to you, they aim to provide a step-by-step development. On the other hand, there is little point in spending time practising something at which you are already very competent.

Some suggestions for tackling exercises:

- Try the first few questions in an exercise. If you find them easy, look for some that seem to offer more of a challenge. Often the last few are harder.

- If you cannot see how to approach a question, ask yourself: 'In what way is this the same as the ones I have already done? What is different about it? Can I adjust the method I have been using for the earlier questions?'

The answer you get to a problem may not look exactly like that given in the solution, because there is often more than one equivalent way of writing down an algebraic expression. However, on closer inspection you may find that the two versions *are* equivalent and your answer is correct. Being able to show that two expressions are equivalent is an important skill in algebra. It is especially useful to be able to write an expression in an equivalent but simpler form.

2.3 Simplifying expressions

Some expressions are more complicated than they need to be. An example is $2x + 5x$, which can be written more simply as $7x$ because two lots of something added to five lots *of the same thing* give seven lots of it in total. This kind of expression, which includes two multiples of the *same* variable added together or subtracted, is useful because it can be simplified. By contrast, the expression $2x + 5y$ cannot be simplified because x and y are different symbols and therefore are not the same thing. So bear in mind that it is not always possible to simplify an expression.

Try a few simplifications for yourself.

Activity 22 Simplification

Simplify the following expressions as far as possible:

(a) $12 \times 2A$ (b) $6B + 7B + 8B$ (c) $4x + 2 \times 3x$ (d) $11y - 5y$

(e) $2p^2 + 5p^2$ (f) $6A + 4B$ (g) $6A + 4B + 2A$ (h) $6A + 7A^2$

The last of the expressions in Activity 22 raises an important issue about simplifying expressions. Although both parts of the expression $6A + 7A^2$ include A, they are not 'two lots of the same thing'. It is worth thinking about this issue more carefully. A mini-expression that is either a number or a multiple of a variable (perhaps to some power) is called a *term*: 9, $6A$, $7A^2$, $4B$ and $2p^2$ are all examples of terms. Terms that are multiples of the same power of the same variable are called *like terms*: $6A$ and $2A$ are like terms, as are $2p^2$ and $5p^2$, but $6A$ and $7A^2$ are not like terms. Remember that, when you simplify an expression that involves different powers, you can combine *like terms* only.

You might like to add this to your Handbook sheet.

Getting like terms together.

The numerical factor in front of a term is often called the *coefficient*; for instance, the coefficient of B in the term $7B$ is 7. So the rule for *adding like terms* is simply *add the coefficients*; thus $7B + 8B = (7 + 8)B = 15B$.

There is one slight oddity: a 'bare' term like x or B^2 does not appear to have a coefficient. But, in fact, it does—its coefficient is the number 1: thus x is the same thing as $1x$. This can be helpful when you have an expression like $x + 2x$; this is the same as $1x + 2x$, which is $3x$.

You might like to add this to your Handbook sheet.

For any particular value(s) of the variable(s), the simplified expression should have the same value as the original expression. A useful check on whether you have simplified correctly is to substitute particular value(s) for the variable(s) in both the original and simplified expressions. The two expressions should give the same result. So, if you put in the value 2 for x in the expression $x + 2x$, you get $2 + 2 \times 2$, which is $2 + 4$ or 6. The value of the simplified expression, $3x$, when x is 2, is 3×2, which is also 6. Whatever value you choose, you should get the same result from the simplified expression as from the original expression.

Activity 23   *More simplification*

Simplify the following expressions as far as possible:

(a) $x + 8x + 3x^2 - 2x^2$ (b) $x + 8x^2 + 3x - 2x^2$

(c) $1 + 2a + 3a^2 + 4a + 5$ (d) $2 \times 3A + 2A^3 + 2^3 A$

(e) $2 \times 5A^2 - 3A \times 2A$ (f) $^-3 \times x^2 + x - 2x \times 4x$

(g) $2z + x^2 + \sqrt{z^2}$ (h) $y^3 + 2y + y^2 - \sqrt{4y^2} - 4$

Check that your simplified expressions take the same values as the originals for a couple of different values of the variables, say, 1 and 2.

You can use your calculator to do this if you wish.

2.4 *Equality*

The equals sign '$=$' is often misused in algebra, so you need to be careful where you place it. It is sometimes *wrongly* used to stand for 'therefore' or 'so' or 'hence'. However, it does not just stand for the word 'equals', but also for phrases like 'is the same thing as' or 'is equivalent to'. In different situations it may have subtly different meanings: it might mean 'is always equal to' or 'is sometimes equal to'. Although the equals sign can be used in a number of contexts, it always means that the expression on one side of the sign is equal to the expression on the other side of it in some respect.

This was discussed in the preparatory audio material.

You may find it helpful to picture the equals sign as an old-fashioned balance: when the 'weights' in each pan are equal numbers or expressions, the pans will balance, as in Figure 6.

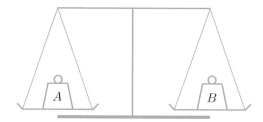

Figure 6   A balance showing that $A = B$.

If you have expressions on either side of an equals sign that do *not* balance, then consider whether something other than 'equals' might be more appropriate: perhaps 'is greater than' or 'produces' would be better.

A common use of the equals sign is *to state the value of a symbol* in a given context. When analysing a 'think of a number' trick, you might say something like 'When N is equal to 7'. The mathematical shorthand way of saying this is 'When $N = 7$'. In this context the equals sign is assigning a value to a variable: the value 7 to the variable N. If the number N is on one side of the balance and 7 is on the other, then the pans will balance (see Figure 7).

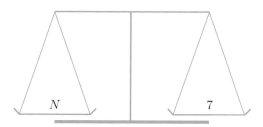

Figure 7   A balance showing that $N = 7$.

A second use of the equals sign is *to state that two expressions are equal.* For example, during the analysis of a number trick you might say that $6N + 10 + 2N + 1$ may be simplified to $8N + 11$ and that hence the two expressions are equal. You can use the equals sign instead of 'may be simplified to' and write $6N + 10 + 2N + 1 = 8N + 11$. In this context the equals sign implies that the two expressions are just different ways of writing the same thing. Another way of putting this is that $6N + 10 + 2N + 1$ and $8N + 11$ always take the same value, whatever the value of N. The two expressions always balance, irrespective of the value of N (see Figure 8 opposite).

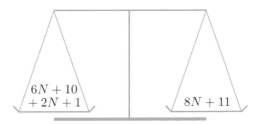

Figure 8   A balance showing that $6N + 10 + 2N + 1 = 8N + 11$ (for all N).

Another context where the equals sign might be used is *in evaluating an expression*. You might write 'When N is equal to 7, then $4N + 5$ has the value 33', but this could be written more concisely as 'When $N = 7$, then $4N + 5 = 33$'. So, with $4N + 5$ on one side and 33 on the other, the scales will balance if N is 7, but not for other values of N (see Figure 9).

With complex expressions, there may be more than one value that balances.

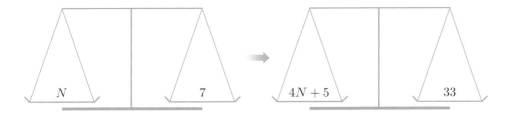

Figure 9   Balances showing that when $N = 7$, $4N + 5 = 33$.

An equals sign is also used in the converse situation, *when solving an equation*. Suppose you want to know 'For what value of N, does $4N + 5 = 17$?' In this case you are seeking the value of N for which $4N + 5$ balances with 17 (see Figure 10). The process of finding this value is called solving the equation. The relevant value of N is 3. Thus $N = 3$ is a solution of the equation $4N + 5 = 17$.

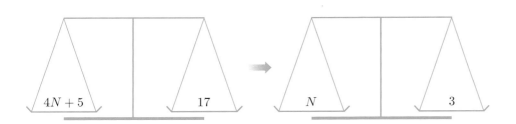

Figure 10   Balances showing that if $4N + 5 = 17$, then $N = 3$.

Activity 24   Using equals signs

Rewrite the following sentences more concisely, by replacing some words/phrases by equals signs where appropriate:

'The expression $4p + 5 + p - 3$ can be simplified to $5p + 2$.'

'When p is equal to 5, the expression $4p + 5 + p - 3$ takes the value 27, and the expression $5p + 2$ also takes the value 27.'

2.5 Brackets

Brackets are often essential in mathematical expressions in order to avoid ambiguity. For instance, suppose that you want to produce a symbolic version of the calculation represented by the following rule:

'Take a number N, add 10 to it, and then multiply the result by 4'.

Adding 10 to N gives $N + 10$, and then multiplying the result by 4 gives $4 \times (N + 10)$. You might have been tempted to write $N + 10 \times 4$ or perhaps $4 \times N + 10$. But both of these would have been wrong, because they do not multiply the whole of the expression $N + 10$ by 4.

You can check this by substituting a number for N: try $N = 1$. In this case, the rule produces: 1 add 10, which gives 11; multiply 11 by 4, which gives 44. However, the expression $N + 10 \times 4$ produces

$$1 + 10 \times 4 = 1 + 40 = 41,$$

while $4 \times N + 10$ produces

$$4 \times 1 + 10 = 4 + 10 = 14.$$

Neither $N + 10 \times 4$ nor $4 \times N + 10$ gives the correct answer. To avoid any ambiguity, use brackets and write $4 \times (N + 10)$ or $(N + 10) \times 4$.

This example illustrates a general principle that when writing an algebraic expression, include in brackets any calculation that should be evaluated *before* what lies outside the brackets is dealt with.

One of the conventions in algebra is that in an expression that involves multiplying a term in brackets, the multiplication sign can be omitted for brevity. Thus $4 \times (N + 10)$ can be written as $4(N + 10)$. Similarly, when writing expressions like 'multiply 2 by $4N$', you can write $2(4N)$. Remember that when you are multiplying several numbers (or symbols representing numbers) together, it does not matter in which order you do the multiplication: $2(4N)$ is $2 \times (4 \times N)$, which is the same as $(2 \times 4) \times N$ (that is, $8N$). Choose whichever is more convenient in the particular situation.

Activity 25 *Brackets*

(a) Evaluate $4(N + 10)$, $4N + 10$ and $N + 10 \times 4$ when $N = 7$.

(b) In the 'think of a number' sequence on the audio, you had to double $2N + 5$ at one stage, and divide $4N + 4$ by 4 at another. Express each of these steps using brackets.

A useful point to note is that an expression enclosed in brackets can be treated as a single entity. So, in $4(N + 10)$, the expression $(N + 10)$ can be regarded as a single entity. This means that $4(N + 10)$ and, say, $2(N + 10)$ can be treated as like terms. Consequently, the expression $4(N + 10) + 2(N + 10)$ may be simplified by adding the coefficients of $(N + 10)$ to give $6(N + 10)$.

Activity 26 *Simplification with brackets*

(a) Express the following expressions without multiplication signs:

  (i) $(m + 3) \times 9$ (ii) $(m + 3) \times (9 + 2 \times m)$.

(b) Simplify the following expressions:

  (i) $2(m + 3) + 3(m + 3)$ (ii) $5(4N + 10) - 2(4N + 10)$

  (iii) $3(9 - 2 \times z) + (9 - 2z) \times 2$.

To sum up, one of the most important uses of brackets is to eliminate ambiguity. It is essential that mathematical formulas are clear and unambiguous; it is therefore crucial for you to be able to use brackets when constructing algebraic expressions. If you have any doubt about how an expression might be interpreted, it is better to put in too many brackets rather than too few. And remember that, to be meaningful, brackets come in pairs!

Multiplying out brackets

Many formulas involve brackets. Sometimes it is more convenient to have the formula with brackets, and sometimes it is not. Therefore it is useful to be able to rewrite a formula that contains brackets, in such a way that the brackets are no longer needed. One important instance where brackets are removed is in the process of *multiplying out brackets*; this process is also referred to as *expanding brackets*. You may remember that this was done first with numbers and then with symbols on the audio band in Section 1. As you may recall, multiplying out $4 \times (5 + 70)$ gives $4 \times 5 + 4 \times 70$: the 4 outside the brackets multiplies *both* of the terms inside the brackets. In the 'think of a number' sequence, multiplying out the brackets in the expression $2(N + 5)$ gives $2 \times N + 2 \times 5$, or $2N + 10$.

You may remember that this is called the distributive property.

Some people find it helpful, when multiplying out brackets, to think of the letters and numbers as representing distances, and the products as representing areas. The idea of products as areas is illustrated in Figure 11, where 6×4 and 6×3 represent the areas of rectangles 6 by 4 and 6 by 3, respectively.

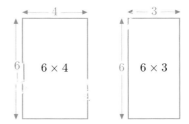

Figure 11   Representing products as areas.

By extending this idea, an expression such as $6 \times (4 + 3)$ can be thought of as representing the area of a rectangle that can be split into two smaller rectangles, as shown in Figure 12. Here the area can be expressed as either $6 \times (4 + 3)$ or $6 \times 4 + 6 \times 3$; the latter form emphasizes that the 6 outside the brackets multiplies both the terms inside the brackets.

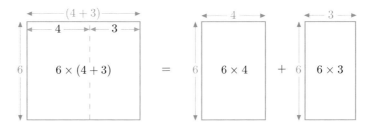

Figure 12   Multiplying out brackets: $6 \times (4 + 3) = 6 \times 4 + 6 \times 3$.

This approach works equally well if some of the terms are letters. Thus $4(N + 10)$ represents the area of a rectangle 4 by $(N + 10)$. This area is made up of two parts: $4 \times N$ and 4×10 (see Figure 13). So

$$4(N + 10) = 4N + 4 \times 10 = 4N + 40.$$

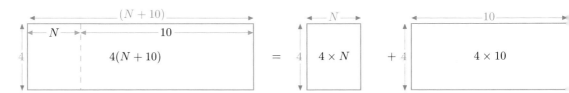

Figure 13   Multiplying out brackets: $4(N + 10) = 4N + 40$.

Example 5 *Removing brackets*

Express $3(X + 3)$ without brackets.

Solution

The key point is that the factor 3 multiplies both of the terms inside the brackets. So

$$3(X + 3) = 3 \times X + 3 \times 3 = 3X + 9.$$

Example 6 *Negatives in brackets*

Express $3(X - 3)$ without brackets.

Solution

Subtracting 3 is the same as adding $^-3$, therefore

$$3(X - 3) = 3(X +^- 3).$$

As the factor 3 multiplies both the terms inside the brackets,

$$3(X - 3) = 3(X +{}^-3) = 3 \times X + 3 \times {}^-3 = 3X + {}^-9 = 3X - 9.$$

With practice you should not need to write down the intermediate steps when multiplying out brackets.

Activity 27 *Doing without brackets*

Express the following without brackets:

(a) $2(m + 3)$ (b) $6(2X + 1)$ (c) $6(x - 2)$ (d) $n(n + 2)$

(e) $6X(2X + 1)$ (f) $5(9 + 2z)$ (g) $3(2m - 6)$ (h) $5p(1 - 7p)$

Now consider expressions with two sets of brackets, such as the product $(n+2)(n+3)$. This represents the area of a rectangle with one side of length $(n+2)$ and the other of length $(n+3)$. Such a rectangle is made up of four smaller rectangles (see Figure 14). The area $(n+2)(n+3)$ must be equal to the sum of these four smaller rectangles. So $(n+2)(n+3) = n \times n + n \times 3 + 2 \times n + 2 \times 3$, which can be simplified to $n^2 + 3n + 2n + 6$. Collecting like terms then gives $n^2 + 5n + 6$. Therefore

$$(n+2)(n+3) = n^2 + 5n + 6.$$

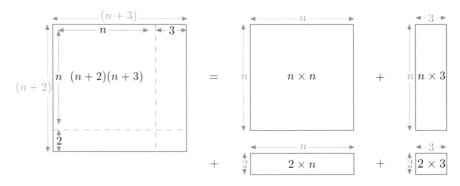

Figure 14   Multiplying out brackets: $(n+2)(n+3)$.

A convenient shorthand way of representing these area diagrams for multiplying out brackets involves drawing a small table, in which the left-hand column represents the terms in the first set of brackets and the top row represents the terms in the second set of brackets. In the case of $(n+2)(n+3)$, the table would be

	n	3
n	n^2	$3n$
2	$2n$	6

Each term in the top row is multiplied by each term in the column to give the four terms in the main body of the table.

From the table, it can be seen that

$$(n+2)(n+3) = n^2 + 3n + 2n + 6$$
$$= n^2 + 5n + 6.$$

This method can be used as a way of multiplying out any brackets, regardless of whether they represent areas or not. For example, to multiply out $(x+7)(2x+3)$, the shorthand version would be

	$2x$	3
x	$2x^2$	$3x$
7	$14x$	21

Hence

$$(x + 7)(2x + 3) = 2x^2 + 3x + 14x + 21 = 2x^2 + 17x + 21.$$

The aim of the area and table methods for multiplying out brackets is to ensure that everything in one set of brackets is multiplied by everything in the other. In order to ensure this has been done, some people prefer to use the acronym FOIL (**F**irst terms, **O**uter terms, **I**nner terms, **L**ast terms). A schematic representation is shown in Figure 15.

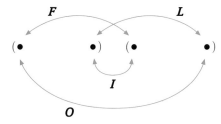

Figure 15   FOIL.

Example 7   *Removing more brackets*

Expand $(2x + 3)(3x + 1)$.

Solution

Everything in the first set of brackets must be multiplied by everything in the second set of brackets, as indicated in Figure 16.

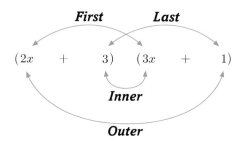

Figure 16   Multiplying out brackets: $(2x + 3)(3x + 1)$.

Multiplying **F**irst terms gives $2x \times 3x$,
multiplying **O**uter terms gives $2x \times 1$,
multiplying **I**nner terms give $3 \times 3x$,
multiplying **L**ast terms gives 3×1.

The table method gives

	$3x$	1
$2x$	$6x^2$	$2x$
3	$9x$	3

Therefore, the multiplication yields four terms: $2x \times 3x$, $2x \times 1$, $3 \times 3x$, 3×1. These simplify to $6x^2$, $2x$, $9x$, 3. Adding these terms together gives $6x^2 + 2x + 9x + 3$.

There are two like terms here involving x, namely $2x$ and $9x$, and these can be added together to give $11x$. So the expansion can be written as

$$(2x + 3)(3x + 1) = 6x^2 + 2x + 9x + 3 = 6x^2 + 11x + 3.$$

In addition to the methods outlined on the preceding pages, there are a number of other ways of multiplying out brackets which may already be familiar to you. Use whichever method you prefer. This unit will generally use the FOIL method, but will also often give the table method.

Activity 28   *More multiplying out*

Recall the advice on doing exercises on p. 37.

Express the following without brackets:

(a) $(y+7)(y+1)$

(b) $(p+2)(p+3)$

(c) $(2x+7)(x+1)$

(d) $(p+2)^2 = (p+2)(p+2)$

(e) $(4p+2)(3p+2)$

(f) $(a+b)(c+d)$

(g) $(2a+b)(c+3d)$

(h) $(3x+4y)(2w+5z)$

When there is a minus sign inside brackets, then, as before, you can treat the subtraction concerned as the equivalent of adding a negative number. Thus $(2x-3)$ can be thought of as $(2x+{}^-3)$.

Example 8   *Including a negative*

Multiply out $(2x-3)(3x+1)$.

Solution

The table method gives

	$3x$	1
$2x$	$6x^2$	$2x$
$^-3$	^-9x	$^-3$

$$
\begin{aligned}
(2x-3)(3x+1) &= (2x+{}^-3)(3x+1) \\
&= 2x \times 3x + 2x \times 1 + {}^-3 \times 3x + {}^-3 \times 1 \\
&= 6x^2 + 2x + {}^-9x + {}^-3 \\
&= 6x^2 + 2x - 9x - 3 \\
&= 6x^2 - 7x - 3.
\end{aligned}
$$

With practice you will probably be able to omit the first couple of steps in manipulations such as that in Example 8.

Activity 29

Expand the following:

(a) $(2x+3)(x-1)$

(b) $(x-1)(4x+2)$

(c) $(2x-1)(3x+1)$

(d) $(x-1)(x+2)$

(e) $(2y+2)(2y-2)$

(f) $(3z+4)(3z-4)$

(g) $(a+b)(a-b)$

(h) $(c-d)(2c+d)$

(i) $(3-x)(2+x)$

Activity 30   *Odd one out*

Find the odd ones out amongst the following (take care to look at all the like terms):

(a) $(m + 3)(m + 5)$,

$5(m + 3) + m^2 + 3m$,

$(5 + m)(3 + m)$,

$m^2 + 8 + 15m$.

(b) $(3x + 1)(x + 2)$,

$2 + x + 3x^2 + 6x$,

$3x(x + 2) + 2(3x + 1)$,

$3x^2 + 7x + 2$.

(c) $(y + 1)(y - 2)$,

$y^2 + y - 2$,

$y^2 - y - 2$,

$y(y - 2) + y - 2$.

Activity 31   *How many ways?*

In five minutes, how many ways can you find of writing the expression $x^2 + 2(x + 3) + 3x$ that are different in form but equivalent in meaning?

You may want to pause here to add notes to your Handbook sheet.

Brackets and negative numbers

'A negative number multiplied by a negative number gives a positive number.'

This claim that 'a minus times a minus is a plus' is, to many people, one of the mysteries of mathematics. There are a number of ways of illustrating why the claim must be true. One very important reason is that mathematics needs to be consistent: when you carry out a calculation in two different ways, you must get the same answer.

Consider, as an example, the calculation $^-4(3 - 1)$. If this calculation is done by evaluating what is in the brackets first, the result is $^-4 \times 2 = ^-8$. However, the calculation could also be done by multiplying out the brackets. If the answer obtained by the second method is different, then something is badly wrong: mathematics is not going to be much use if it is not consistent!

If you are not sure about why $^-4 \times 2 = ^-8$, you may find it helpful to refer to the *Preparatory Resource Books*, Module 1.

So look at what happens when the brackets are multiplied out:

$$^-4(3 + ^-1) = ^-4 \times 3 + ^-4 \times ^-1.$$

The first term, $^-4 \times 3$, is $^-12$. But what is $^-4 \times ^-1$? Well, there are two plausible options: either $+4$ or $^-4$. If $^-4 \times ^-1$ were $^-4$, then the expression $^-4(3 + ^-1)$ would be $^-12 + ^-4 = ^-16$, which is not the correct answer! If $^-4 \times ^-1$ were $+4$, then the expression would be $^-12 + 4$, which equals $^-8$, as required.

Although this is just one example, the same principle would apply whatever negative numbers were used—it is quite general. Therefore, if mathematics is to be consistent, then the *product of two negative numbers must be positive*.

The following example and activity will give you experience of multiplying two negative numbers together.

Example 9   *Multiplying out brackets involving negatives*

Expand $(2x - 3)(3x - 1)$.

Solution

Rewrite this expression as $(2x + {}^-3)(3x + {}^-1)$. Then multiply it out. Everything in the first set of brackets must be multiplied by everything in the second set of brackets, as indicated in Figure 17.

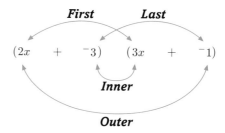

Figure 17   Multiplying out
$(2x + {}^-3)(3x + {}^-1)$.

Using the fact that the product of two negative numbers is positive, ${}^-3 \times {}^-1 = +3$.

This results in four terms: $2x \times 3x$, $2x \times {}^-1$, ${}^-3 \times 3x$, ${}^-3 \times {}^-1$. These simplify to $6x^2$, ${}^-2x$, ${}^-9x$, 3. So adding these terms together gives $6x^2 + {}^-2x + {}^-9x + 3$. Since there are two terms involving x, namely ${}^-2x$ and ${}^-9x$, these can be collected together. Thus

$$(2x - 3)(3x - 1) = 6x^2 + {}^-2x + {}^-9x + 3$$
$$= 6x^2 + {}^-11x + 3, \quad \text{or} \quad 6x^2 - 11x + 3.$$

The table method gives

	$3x$	$^-1$
$2x$	$6x^2$	^-2x
$^-3$	^-9x	3

As in earlier examples, you may find that you do not need to do all the intermediate steps; in particular, you may not need to replace -3 by $+{}^-3$. You may simply be able to write the expansion as

$$(2x - 3)(3x - 1) = 6x^2 - 2x - 9x + 3$$
$$= 6x^2 - 11x + 3.$$

Activity 32

Expand the following:

(a) $(2x - 3)(x - 1)$ (b) $(3x - 1)(4x - 2)$ (c) $(1 - x)(4x - 2)$

(d) $(x - 1)^2$, which is $(x - 1)(x - 1)$ (e) $(x - 1)(x + 1)$

Dividing a number into brackets

In general terms, dividing by a number is the same as multiplying by its reciprocal. Thus, dividing by 2 is the same as multiplying by $\frac{1}{2}$. This means that the process of multiplying out brackets can be adapted to deal with the case of brackets that are being divided rather than being multiplied.

If you are unsure of this, then refer back to Module 1 of the Preparatory Resource Books.

For example, $(10x + 6) \div 2$ is the same as $(10x + 6) \times \frac{1}{2}$. So

$$(10x + 6) \div 2 = (10x + 6) \times \tfrac{1}{2}$$
$$= 10x \times \tfrac{1}{2} + 6 \times \tfrac{1}{2}$$
$$= 10x \div 2 + 6 \div 2$$
$$= 5x + 3.$$

Note that *each* term in the brackets must be divided by the 2.

Activity 33   *Removing brackets involving division*

Express the following without brackets:

(a) $(4N + 8) \div 4$ (b) $(6m + 3) \div 3$

(c) $5(4 - 2z) \div 2$ (d) $(6k - 12) \div {}^{-}3$

The next activity gives you practice in the various uses of brackets.

Activity 34   *Incorrect brackets*

Which of the following mathematical statements involving brackets are correct and which incorrect? Correct the incorrect statements.

(a) $5 \times (N + 2) = 5N + 10$

(b) $6 \times (2N + 1) = 12N + 1$

(c) $2(N + 1) + 3(N + 1) = 5(N + 1)$

(d) $(N + 2)(N + 2) = N^2 + 4N + 4$

(e) $N + 3(N + 3) = N^2 + 6N + 9$

(f) $(X + 2(X + 3) = X^2 + 5X + 6$

(g) $X + 2)(X + 4) = X^2 + 6X + 8$

(h) $(X - 2)(2X + 2) = X^2 - 2X - 4$

(i) $(2X - 4) \div 2 = X - 2$

(j) $(4X - 2 \div 2 = 2X - 1$

(k) $3X + 6) \div 3 = X + 2$

Algebraic fractions

When writing algebraic expressions there is an important difference between multiplication and division. Unlike the multiplication sign, the division sign cannot be dispensed with. Clearly, $(4N + 8) \div 4$ is not the same as $(4N + 8)4$, because in the latter expression the 4 multiplies what is in the brackets rather than divides it.

As in arithmetic, there are alternatives to the use of the division sign. These alternatives are based on various different ways of writing fractions. Instead of $(4N + 8) \div 4$, you can write any of the following:

$$(4N + 8)/4,$$

$$\tfrac{1}{4}(4N + 8),$$

In $\dfrac{4N + 8}{4}$, the division sign effectively acts as a set of brackets.

$$\frac{(4N + 8)}{4}, \quad \text{or} \quad \frac{4N + 8}{4}.$$

All of these expressions simplify to $N + 2$.

Notice the importance of using brackets in the first two of these alternatives. In the expression $(4N + 8)/4$, if you leave out the brackets you get $4N + 8/4$, which is $4N + 2$ and is incorrect. If you omit the brackets in the case of $\tfrac{1}{4}(4N + 8)$, you get $\tfrac{1}{4}4N + 8 = N + 8$, which is also incorrect.

These alternative ways of writing fractions also apply to fractions involving algebraic expressions in both the numerator (top) and the denominator (bottom); such expressions are called *algebraic fractions*, or *quotients*. For example, $(4N + 8)$ divided by $(4N^2 + 1)$ can be written as

$$(4N + 8)/(4N^2 + 1),$$

or

$$\frac{1}{4N^2 + 1}(4N + 8),$$

or

In your written work, you are recommended to write algebraic fractions in either of the last two of these forms.

$$\frac{(4N + 8)}{(4N^2 + 1)},$$

or

$$\frac{4N + 8}{4N^2 + 1}.$$

In the last of these expressions, leaving out the brackets does not cause any ambiguity; there is nothing else that the expression could mean except 'evaluate $4N + 8$ and $4N^2 + 1$, and then divide the former by the latter'. However, it is essential to include the brackets when using the slash (/) version, as exemplified by the first expression.

All of these expressions, except the second one, are read as '$4N + 8$, all over $4N^2 + 1$'; the second expression is read as 'one over $4N^2$ plus 1, multiplied by (pause) $4N$ plus 8 in brackets'.

The 'pause' is to show that you are no longer on the bottom line.

You may see algebraic fractions written in any of these forms. The version with the slash is sometimes used in text to avoid taking up two lines! The 'built up' fractions are commonly used in expressions that are displayed on separate lines.

Example 10   *Evaluating an algebraic fraction*

Evaluate (a) $\dfrac{2X + 1}{3X - 1}$ and (b) $(6X + 3)/(4X - 1)$ when $X = 1$, and when $X = 4$.

Solution

(a) When $X = 1$, $2X + 1 = 3$ and $3X - 1 = 2$. So

$$\frac{2X + 1}{3X - 1} = \frac{3}{2}, \quad \text{or} \quad 1.5.$$

When $X = 4$,

$$\frac{2X + 1}{3X - 1} = \frac{9}{11}.$$

(b) When $X = 1$, $(6X + 3) = 6 + 3 = 9$ and $(4X - 1) = 4 - 1 = 3$. So

$$(6X + 3)/(4X - 1) = 9/3 = 3.$$

When $X = 4$, $(6X + 3) = 27$ and $(4X - 1) = 15$. So

$$(6X + 3)/(4X - 1) = 27/15 = \tfrac{9}{5}, \quad \text{or} \quad 1.8.$$

Activity 35

(a) Evaluate $(4X - 7)/(2X - 3)$ when $X = 2$, and when $X = 1$.

(b) Evaluate $\dfrac{6X - 2}{3X + 1}$ when $X = 7$.

When dividing an expression in brackets by a number, you saw how you could multiply by the reciprocal of the number instead. Similarly, when dividing an expression in brackets by another algebraic expression, you can multiply by the reciprocal of the second expression. For instance,

See p. 51.

$$(4X - 3) \div \frac{(X - 1)}{3} = (4X - 3) \times \frac{3}{(X - 1)}$$
$$= \frac{3(4X - 3)}{(X - 1)}.$$

Think of a number revisited

Some of the ideas that have been presented in this section are now illustrated by taking another look at the 'think of a number' sequence from Section 1.1.

The relevant instructions are shown in Figure 18, which is similar to Frame 18 on page 14. Once again, N is used to stand for the number you first thought of. However, now, instead of simplifying the algebraic expressions at each step, they are written out using brackets.

Instruction	*Algebraic expression*
Think of a number	N
Double it	$2N$
Add 5	$2N + 5$
Double the result	$2(2N + 5)$
Subtract 6	$2(2N + 5) - 6$
Divide by 4	$\dfrac{2(2N + 5) - 6}{4}$
Take away the number you first thought of	$\dfrac{2(2N + 5) - 6}{4} - N$

Figure 18   'Think of a number' revisited.

The sequence of instructions produces the final expression

$$\frac{2(2N + 5) - 6}{4} - N.$$

This looks quite complicated. But, of course, the reason that the 'trick' works is that the expression reduces to something that is not complicated at all. Some algebraic manipulation shows why.

The expression can be simplified in a number of different ways. One way is to carry out the steps below:

(i) Multiply out the expression involving brackets, namely $2(2N + 5)$, and substitute the result in the expression itself to obtain

$$\frac{4N + 10 - 6}{4} - N.$$

(ii) Collect like terms (here, the numbers 10 and $^-6$) to get

$$\frac{4N + 4}{4} - N.$$

(iii) Divide out the fraction to get

$$N + 1 - N.$$

(iv) Collect like terms again (the Ns this time), and obtain the answer 1.

This process of simplification would normally be written out as follows:

$$\frac{2(2N+5)-6}{4} - N = \frac{4N+10-6}{4} - N$$
$$= \frac{4N+4}{4} - N$$
$$= N + 1 - N$$
$$= 1.$$

So the basis of the 'trick' is to use an expression that looks complicated, but actually unravels (when you simplify it) to a number that is independent of the number you first thought of.

If you have ambitions to be known as a mind-reader, you might like to construct a 'think of a number' trick for yourself. It is usually better to simplify the expression after each step (as in Frame 18), rather than leave all the algebraic manipulation until the end.

Activity 36   *Think of a number again*

(a) Think of a positive number; subtract 1; square the result; add twice the number you first thought of; subtract 1; take the positive square root; take away the number you first thought of, and your answer is 0. Explain why this is so algebraically.

(b) These are some steps in an incomplete 'think of a number' sequence: Think of a number, X; triple it; add 10; double the result; subtract 8.

Work out an expression for the outcome of these steps and simplify it. Now add further steps to complete the sequence, so that it ends 'and your answer is 2'.

(c) Complete the 'think of a number' sequence that begins: 'Think of a number; quadruple it' and ends 'take away the number you first thought of, and your answer is 3'.

Activity 37   *The language of algebra*

Before completing this section, review what you have covered. Add to your notes or to the Handbook sheet, so that you have a record of the new ideas that you have met. This should give you a summary of the language of algebra. Include some examples if you find them helpful.

Make sure you understand the meaning of the following: algebraic expression; variable; evaluating an expression; substituting for a variable; value of a variable; value of an expression; simplifying an expression; multiplying out brackets; coefficient; term. Also make sure you are confident about using equals signs and brackets.

Now consider your progress by referring to the section outcomes below. Do you feel confident about using the skills and techniques introduced in this section, or do you need more practice? Would doing some of the exercises in the *Resource Book* be useful?

Outcomes

After studying this section, you should be able to:

◇ recognize whether or not collections of symbols are valid algebraic expressions (Activity 19 and 34);

◇ understand the meanings of the terms: 'substitute', 'evaluate', 'value' (Activities 17 and 18);

◇ evaluate an algebraic expression by substituting a value for the variable (Activities 17, 18, 24, 25 and 35);

◇ multiply or divide an algebraic expression by a number (Activities 27 and 33);

◇ multiply out expressions in brackets, including those that incorporate symbols and both positive and negative numbers (Activities 27–32, 34 and 36);

◇ simplify an algebraic expression (Activities 20, 22, 23, 26, 31 and 36);

◇ consider your progress in learning algebra (Activities 21 and 37).

3 Algebra and your calculator: functions

Aims   This section aims to teach you how to enter formulas into your calculator, and how to use your calculator to evaluate them.   ◇

3.1 More about mathematical functions

As described in Section 1, a mathematical function takes input numbers, processes them according to some rule, and produces output numbers. Many of the keys on a calculator represent mathematical functions. For example, the $\boxed{x^2}$ key allows you to carry out the mathematical function of squaring a number: in the course calculator you enter a number, press the $\boxed{x^2}$ key and $\boxed{\text{ENTER}}$, and obtain the square of that number. The *input* was the number you entered, the *rule* was 'square the number', and the *output* was the square of the number. The general idea is shown in Figure 19.

Some of the so-called 'function keys' on a calculator do not represent mathematical functions but are used to change the settings (such as the number of decimal places in the display) or to access other screens. In these cases, 'function' simply means 'what the key does', as in 'first function' etc. in the *Calculator Book*.

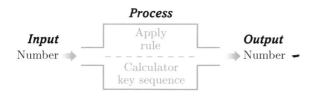

Figure 19   A representation of a function.

On the calculator the operation keys $+$, $\times$, $-$ and $\div$ can be used as part of a key sequence to represent a mathematical function. For example, the function $K = 1.61 \times M$, which is used to convert miles to kilometres, can be represented by the key sequence $\boxed{\times}$ 1.61 $\boxed{\text{ENTER}}$. If you input a number (in miles), press the $\boxed{\times}$ key followed by 1.61 and then $\boxed{\text{ENTER}}$, you get the output number (in kilometres), as Figure 20 indicates.

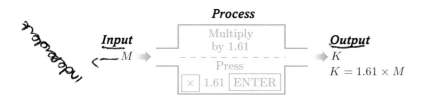

Figure 20   The function $K = 1.61 \times M$.

In Section 7.2 of the *Calculator Book*, you entered the mile/kilometre conversion function into your calculator. Recall that you entered the function as $Y = 1.61 \times X$, not as $K = 1.61 \times M$. This change of letters makes no essential difference: the substance of the formula lies in the algebraic relationship it expresses, and this is not affected by the substitution of X for M and Y for K.

With X representing the input variable, and Y representing the output variable, functions take the form:

$Y = $ an expression involving X.

An 'expression involving X' may also be referred to as 'a function of X'.

The variable Y is called the *dependent variable*, because it depends on X (indeed, the formula shows exactly how it depends on X), and X is called the *independent variable*. Therefore, in the function $K = 1.61 \times M$, the distance in miles, M, is the independent variable, and the distance in kilometres, K, the dependent variable.

When using K and M, the letters help you to remember the meanings of the symbols. These clues are lost when using X and Y.

Another common choice of letters for the independent and dependent variables is lower case x and y, respectively.

To summarize, a formula (with one input) is a function if it is written in the form:

dependent variable $=$ an expression involving the independent variable

A common choice of letters to stand for the variables are:

X for the independent variable (that is, the input for the function),

Y for the dependent variable (that is, the output of the function).

Then

$Y = $ a function of X,

which is sometimes abbreviated to $Y = f(X)$.

With x and y as the choice of letters, '$y = $ a function of x' is written $y = f(x)$.

3.2   *Using the calculator with formulas*

You will now learn how to use your calculator to enter and store formulas that can be represented by mathematical functions. Once a function is stored, you can do several different things with it. For instance, you can get the calculator to evaluate it for any chosen value of the variable (that is, for any number you choose), or to draw up a table giving the values of the function for different values of the variable.

Work through Sections 8.1, 8.2 and 8.3 of Chapter 8 of the Calculator Book.

Activity 38   *Handbook functions*

Add to your notes on mathematical functions from Activity 7, explaining how to enter and use mathematical functions on your calculator.

Outcomes

After studying this section, you should be able to:

◇   explain the term *mathematical function* in your own words (Activity 38);

◇   understand the difference between independent and dependent variables (Activity 38);

◇   enter and store functions in your calculator;

◇   use your calculator to evaluate a stored function for particular values of the independent variable;

◇   use your calculator to produce a table of values for a function.

4   A change of subject

Aims   This section aims to teach you how to reverse formulas by using the idea of the inverse of a mathematical function; this involves identifying the operations by which the function is built up, and then reversing those operations.   ◇

4.1   Doing and undoing

Recall that the two formulas for converting between distances measured in miles and in kilometres are

$$K = 1.61 \times M \quad \text{and} \quad M = K \div 1.61.$$

Recall in this context the doing–undoing ideas and diagrams in Chapter 1.7 of the *Calculator Book*. Multiplication and division undo one another, as do addition and subtraction.

These two formulas are mathematical functions, and they specify a pair of reverse processes, or inverse functions. What one function 'does', the other one 'undoes'. If 'multiply by 1.61' is doing, then 'divide by 1.61' is undoing. If you start with a distance measured as 20 miles and use the formula $K = 1.61 \times M$ with $M = 20$, you get $K = 32.2$, so the distance is also 32.2 km. If you put $K = 32.2$ into the formula $M = K \div 1.61$, then it undoes the process, as shown in Figure 21, and you get back to $M = 20$, that is, a distance of 20 miles.

Doing process

| **Input** | Multiply by 1.61 | **Output** |
| 20 ➡ | | ➡ 32.2 |

Cancel each other out

Undoing process

| **Output** | Divide by 1.61 | **Input** |
| 20 ⬅ | | ⬅ 32.2 |

Figure 21   Doing and undoing.

In the formula $K = 1.61 \times M$, K is the *subject* of the formula, and in the formula $M = K \div 1.61$, M is the subject. When you change $K = 1.61 \times M$ into the form $M = K \div 1.61$, you are said to *change the subject* of the formula.

This example illustrates a method of changing the subject of a formula. For convenience, it can be referred to as the *'doing and undoing'* method.

Activity 39 *Undoing Kelvin*

For many scientific purposes, temperatures are measured on the absolute scale known as the Kelvin scale.

The formula for converting a temperature of $c\,°$Celsius to k on the Kelvin scale is $k = c + 273$.

If this is a 'doing' formula, what is the corresponding 'undoing' formula?

The Kelvin scale was mentioned in *Unit 7*.

So far, the formulas in this section have involved only one operation. When a formula has *more* than one operation, there will be more than one step in the doing process. This makes the undoing process a little more complicated: the steps have to be undone successively but in the *reverse order*.

You have previously met formulas involving two operations when converting between Celsius and Fahrenheit measures of temperature. These formulas are

See Section 1.2.

$$c = (f - 32) \div 1.8 \quad \text{and} \quad f = 1.8c + 32,$$

where c is the temperature on the Centigrade scale, and f is the same temperature on the Fahrenheit scale. In the first formula, c is the subject, whereas in the second formula, f is the subject.

Starting with a temperature of $212\,°$F and using the first formula to convert it to the Celsius scale, you get $100\,°$C. If you then use the second formula to convert $100\,°$C back to the Fahrenheit scale, you end up with $212\,°$F again.

The doing steps in the formula

$$c = (f - 32) \div 1.8$$

are 'subtract 32' and 'divide by 1.8', and the corresponding undoing steps are 'multiply by 1.8' and 'add 32' (see Figure 22 overleaf). So to make f the subject of the formula $c = (f - 32) \div 1.8$, these undoing steps must be applied to c in the reverse order to get

$$f = 1.8c + 32.$$

Notice that it is crucial to undo first what you did last.

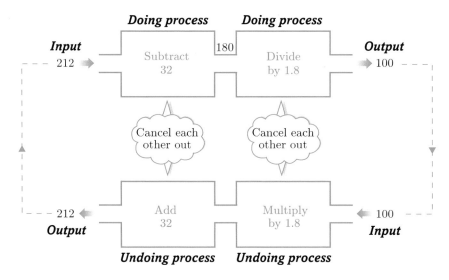

Figure 22   Fahrenheit to Celsius and back.

Reversing the order of the operations in this way may not seem intuitive, but bear in mind that in everyday life it is quite common to reverse the order of operations when you want to undo something that you have done. For instance, you put on your underclothes before putting on your outer clothes, but you take off your outer clothes before your underclothes. You put on your socks or tights before your shoes, but you take off your shoes before you remove your socks or tights.

Occasionally in algebra, as in other areas of life, you can undo two operations together. But it is safer, especially if you are not too confident, *always to undo operations in the reverse order to that in which they were done.*

4.2   Rearranging formulas

The formulas in Section 4.1 involved only one or two operations, but the same principle applies if there are more operations, although then it can be harder to keep track of the doing and undoing steps.

Although the 'doing and undoing' strategy works well for simple formulas, more complicated formulas often require a different method. In this alternative method, as with the 'doing and undoing' method, you first need to see how the formula is built up and list the relevant doing and undoing steps. You then apply the appropriate undoing step to the expressions on *both sides* of the equals sign; this preserves the equality and produces a different, but equivalent, formula. Usually, several of these steps are needed, each producing an equivalent formula, until the required variable is isolated on one side of the formula.

Before you use this method on more complicated formulas, look at how it works when changing the subject of $k = c + 273$.

In order to get c on its own on the right-hand side of the formula, you need to remove the '+273' from that side. To do this, subtract 273. This operation has to be applied to *both sides of the formula*, in order to preserve the equality. So starting with

$$k = c + 273,$$

subtract 273 from *both sides*:

$$k - 273 = c + 273 - 273.$$

On simplifying,

$$k - 273 = c.$$

Turning this round gives the required reverse conversion formula:

$$c = k - 273.$$

This method of rearranging formulas can be called the *'doing the same to both sides'* method.

Always remember that if you add (or subtract) a number in an expression on one side of an equals sign, you must add (or subtract) the same number in the expression on the other side of the sign, so that the expressions on the two sides remain equal and the equation balances.

Example 11

The formula that converts the measure of a volume, G, in British gallons to its measure, L, in litres is

$$L = 4.55 \times G.$$

Change the subject of this formula to G (that is, obtain a formula that converts the measure of a volume, L, in litres to its measure, G, in gallons) by manipulating the formula to get G on its own.

Solution

To get G on its own on the right-hand side of the equals sign in place of $4.55 \times G$, you need to undo the multiplication by 4.55; this necessitates dividing by 4.55. In order to keep things balanced, the expressions on *both sides of the equals sign* must be divided by 4.55. Sometimes this is referred to as *dividing through* by 4.55. This gives

$$\frac{L}{4.55} = \frac{4.55G}{4.55},$$

which simplifies to

$$\frac{L}{4.55} = G.$$

It is usual to put the subject of the formula on the left, so

$$G = L \div 4.55.$$

Activity 40   *Pounds to kilos*

The mass, J, in kilograms of an object is given in terms of its mass, P, in pounds by the formula $J = 0.45P$. What is the formula that represents the mass in pounds in terms of the mass in kilograms?

Activity 41   *Ann's family*

(a) Ann was 25 years old when her son Kim was born. So his age, y years, is given in terms of her age, x years, by the formula

$$y = x - 25.$$

Now retell the story from Kim's point of view. He says: 'When I was born, my mother was 25. So if I want to work out her age, knowing mine, I must ...'. Complete the sentence.

(b) Part (a) gave the personal version of the question. The mathematical version is: Rearrange the formula $y = x - 25$ to make x its subject. Now perform this rearrangement.

A formula expresses a relationship between variables. Rearranging a formula does not change the relationship, but shows it from a different point of view. For example, because M is the subject of $M = K \div 1.61$, this formula can be used to calculate a distance in miles if you know the distance in kilometres. This formula expresses the relationship between the distances from the point of view of, say, a British person visiting a country where distances are measured in kilometres. The formula $K = 1.61 \times M$ expresses the same relationship, but from the point of view of somebody who comes from a country where distances are measured in kilometres, but is visiting the UK, where miles are used.

To generalize, rearranging a formula usually involves changing its subject from one variable to another. The methods used can be summarized as follows:

- To rearrange a formula of the type $y = x + a$ to make x its subject, subtract a from *both sides*:

  $$y - a = x + a - a.$$

  Simplified, this is

  $$y - a = x, \quad \text{or} \quad x = y - a.$$

- To rearrange a formula of the type $y = bx$ to make x its subject, divide *both sides* by the constant b:

  $$y \div b = bx \div b.$$

  This gives

  $$y \div b = x, \quad \text{or} \quad x = y \div b.$$

In both of these cases, rearranging the formula involves undoing, or reversing, a single operation. To undo the addition of a, subtract a, and to undo the multiplication by b, divide by b. Similarly, in an equation of the form $y = x - c$, the subtraction of c is undone by adding c, and in $y = x \div d$, the division by d is undone by multiplying by d. Remember that addition and subtraction are reverse operations, as are multiplication and division.

In *Unit 7* a method was given for remembering the three formulas connecting speed s, distance travelled d and time taken t:

$$d = st, \quad s = \frac{d}{t}, \quad t = \frac{d}{s}.$$

These formulas all express the same relationship, so they are all versions of a single formula but with the subject changed. This means that you need only remember one of the formulas: the others can be obtained by changing the subject.

Which formula do you find easiest to remember? Some people remember $d = st$ because it does not involve division; others find $s = d/t$ easiest because it corresponds to how speeds are measured (as in km/h).

For instance, suppose you want to rearrange $d = st$ in order to make s the subject, you have to undo multiplying by t, so as to get s on its own. This is done by dividing both sides by t, which gives

$$\frac{d}{t} = \frac{st}{t},$$

so

$$\frac{d}{t} = s, \quad \text{or} \quad s = \frac{d}{t}.$$

Activity 42 *Changing speed*

(a) Change $d = st$ to make t the subject.

(b) Change $s = \dfrac{d}{t}$ to make d the subject.

The two methods for changing the subject of a formula, the 'doing and undoing' method and the 'doing the same to both sides' method, will now each be used to change the subject of a formula involving three operations.

Example 12 *Kelvin to Fahrenheit*

A formula for converting from the Kelvin temperature scale to the Fahrenheit scale is $f = 1.8(k - 273) + 32$. Change the subject of the formula to k by using each method in turn.

Solution

With either method, you first need to list the doing and undoing steps that are involved in building up the expression $1.8(k - 273) = 32$ from k.

The first *doing* step is 'subtract 273' (the expression in brackets is dealt with first); this gives $(k - 273)$.

The second *doing* step is 'multiply by 1.8', giving $1.8(k - 273)$.

The final *doing* step is 'add 32', giving $f = 1.8(k - 273) + 32$.

The undoing steps have to be carried out in the reverse order.

The first *undoing* step is 'subtract 32'.

The second *undoing* step is 'divide by 1.8'.

The final *undoing* step is 'add 273'.

For the 'doing and undoing' method, Figure 23 shows the three doing steps starting with k, and also the undoing steps starting with f. The formula is shown after each step. The final formula, with k as the subject, is

$$k = (f - 32) \div 1.8 + 273.$$

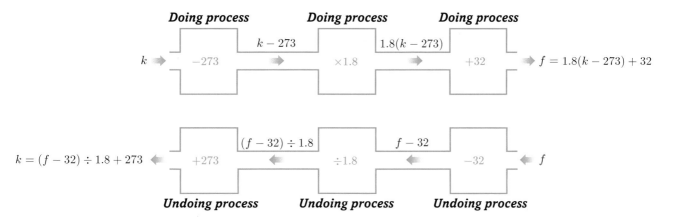

Figure 23   Doing and undoing $f = 1.8(k - 273) + 32$.

With the 'doing the same to both sides' method, the undoing steps are applied successively to *both sides* of the formula. The steps are 'subtract 32', 'divide by 1.8', and then 'add 273'. The details of this rearrangement are as follows.

Start with

$$f = 1.8(k - 273) + 32$$

and subtract 32 from both sides, so

$$f - 32 = 1.8(k - 273) + 32 - 32,$$

or

$$f - 32 = 1.8(k - 273).$$

Then divide both sides by 1.8:

$$(f - 32) \div 1.8 = 1.8(k - 273) \div 1.8.$$

This gives

$$(f - 32) \div 1.8 = k - 273.$$

Finally, add 273 to both sides:

$$(f - 32) \div 1.8 + 273 = k - 273 + 273.$$

This gives

$$(f - 32) \div 1.8 + 273 = k.$$

Hence the rearranged formula is

$$k = (f - 32) \div 1.8 + 273,$$

which can also be written as

$$k = \frac{f - 32}{1.8} + 273.$$

Although it is possible to apply the 'doing and undoing' method directly to such a complicated formula, it is usually safer to use the slightly longer method of 'doing the same to both sides'.

Activity 43   Back again

List the doing and undoing steps involved in rearranging the following formula to make f the subject:

$$c = (f - 32) \div 1.8.$$

Then perform the rearrangement, using the 'doing the same to both sides' method.

The manipulation of the temperature conversion formulas in Example 12 and Activity 43 has demonstrated the principles of how to rearrange complicated formulas that involve a number of operations. In essence, a sequence of steps is performed in the *reverse order* to that in which they were done in the original formula, thereby *undoing* one operation at a time; each step is applied to *both sides* of the equation to preserve equality. Another illustration of this process is provided by the next example.

Example 13   Hedge formulas

Is a hedge mentioned in a historic document the same one that exists today? To test this, you need to know how many species to expect in the hedge. Hooper's rule, which you met in Section 1.1, gives the age, A, of an English hedge in years when the number of species, N, on average, in a 30-yard section of the hedge is known. That formula, $A = 110N + 30$, has A as its subject.

Rearrange the formula to make N the subject.

Solution

First, look at the right-hand side of the given formula

$$A = 110N + 30$$

to see what operations have been 'done' to N, and in what order. The operations must have been: 'multiply by 110', and then 'add 30'.

To make N the subject of the formula, these two operations must be undone in the reverse order.

The first step is to undo 'add 30'. So 'subtract 30' from both sides and get

$$A - 30 = 110N.$$

The second step is to undo 'multiply by 110' by dividing both sides by 110:

$$\frac{A - 30}{110} = N.$$

This formula is the same as

$$N = \frac{A - 30}{110}.$$

Note that this could also be written as

$$N = \frac{1}{110}(A - 30), \quad \text{or} \quad N = (A - 30) \div 110.$$

You can now work out how many species, N, to expect in a hedge of a given age, and then visit the hedge to check.

Remember that, in changing the subject of a formula, the order in which you carry out the operations makes a great deal of difference. In the last example, the doing stage was 'multiply by 110' and then 'add 30', giving $A = 110N + 30$. If you do not reverse the order of the operations when undoing, then you do not end up with the formula in the form required. This is what would happen.

Starting with

$$A = 110N + 30,$$

divide both sides by 110:

$$\frac{A}{110} = N + \frac{30}{110}.$$

Then subtract 30 from both sides:

$$\frac{A}{110} - 30 = N + \frac{30}{110} - 30.$$

This gives

$$\frac{A}{110} - 30 = N - 29\tfrac{8}{11}.$$

In this form, N has not been isolated as the subject of the formula and further manipulation would be needed.

If you have problems in rearranging a formula, it is always worth checking that you have carried out the steps in the right order.

So far, the operations to be undone have involved only addition, subtraction, multiplication and division. Some other operations that may need to be undone when rearranging formulas are summarized below.

Doing	Undoing
Square root ($\sqrt{x}$)	Square (x^2)
Cube root ($\sqrt[3]{x}$)	Cube (x^3)
Square (x^2)	Square root ($\sqrt{x}$)
Cube (x^3)	Cube root ($\sqrt[3]{x}$)
Reciprocal (x^{-1})	Reciprocal (x^{-1})

Note that you cannot always undo squaring uniquely if the variable can take both positive and negative values. In that case, $^-\sqrt{x}$ and $\sqrt{x}$ are both possible results.

The reciprocal may be given in the form $y = \frac{1}{x}$ rather than $y = x^{-1}$. The doing operation is expressed as 'take the reciprocal of', and the undoing operation is also 'take the reciprocal of'. Thus, reciprocals occur in pairs: 2 is the reciprocal of $\frac{1}{2}$, and $\frac{1}{2}$ is the reciprocal of 2. Similarly, $\frac{2}{3}$ and $\frac{3}{2}$ are reciprocals of each other.

Example 14   *Rearranging reciprocals*

Make x the subject of $y = \dfrac{1}{x}$.

Solution

The formula $y = \dfrac{1}{x}$ can be written as

$$\frac{y}{1} = \frac{1}{x}.$$

Taking the reciprocals of both sides gives

$$\frac{1}{y} = \frac{x}{1}, \quad \text{or} \quad \frac{1}{y} = x.$$

So

$$x = \frac{1}{y}.$$

Example 15   *More rearrangements*

Make x the subject of the following formulas:

(a) $y = \dfrac{x}{2} - 3$

(b) $y = 4 + \dfrac{2x}{3}$

(c) $y = \dfrac{3}{2x + 2}.$

Solution

(a) Here the doing steps are 'divide by 2' and 'subtract 3'.

The corresponding undoing steps are 'add 3' and 'multiply by 2'.

Therefore to make x the subject of

$$y = \frac{x}{2} - 3,$$

first add 3 to both sides:

$$y + 3 = \frac{x}{2}.$$

Then multiply both sides by 2:

$$2y + 6 = x.$$

Hence

$$x = 2y + 6.$$

(b) The doing steps are 'multiply by 2', 'divide by 3' and 'add 4'.

The corresponding undoing steps are 'subtract 4', 'multiply by 3' and 'divide by 2'.

Therefore to make x the subject of

$$y = 4 + \frac{2x}{3},$$

first subtract 4 from both sides:

$$y - 4 = \frac{2x}{3}.$$

Then multiply both sides by 3:

$$3y - 12 = 2x.$$

Finally, divide both sides by 2:

$$\frac{3}{2}y - 6 = x.$$

Hence

$$x = \frac{3}{2}y - 6.$$

(c) The doing steps are 'multiply by 2', 'add 2', 'take the reciprocal' and 'multiply by 3'.

The corresponding undoing steps are 'divide by 3', 'take the reciprocal', 'subtract 2' and 'divide by 2'.

Therefore to make x the subject of

$$y = \frac{3}{2x + 2},$$

first divide both sides by 3:

$$\frac{y}{3} = \frac{1}{2x + 2}.$$

Then take the reciprocals of both sides:

$$\frac{3}{y} = \frac{2x+2}{1} = 2x+2.$$

Next, subtract 2 from both sides:

$$\frac{3}{y} - 2 = 2x.$$

Finally, divide both sides by 2:

$$\frac{3}{2y} - 1 = x.$$

Hence

$$x = \frac{3}{2y} - 1.$$

Activity 44   A different subject

Recall the advice about doing sets of exercises given on page 37.

In each of the following, make x the subject of the formula:

(a) $y = 2x + 1$

(b) $y = \dfrac{1}{2}x - 1$

(c) $y = 2(x + 1)$

(d) $y = \dfrac{x}{5}$

(e) $y = \dfrac{x}{5} + 1$

(f) $y = \dfrac{5x - 4}{3}$

(g) $y = \dfrac{3(x + 1)}{4}$

(h) $y = \dfrac{2}{3}(x + 2)$

(i) $y = \sqrt{x}$

(j) $y = \dfrac{21}{x}$

(k) $y = 6 + \dfrac{x}{20}$

(l) $y = 2x^3$

(m) $y = \dfrac{21}{x + 1}$

Activity 45   Detective work

In the detective story *The Ten Nailers* by Dorothy L. Singers set in the 1930s, there is an important clue to the crime; it concerns the hire of a motorbike by one of the suspects. The detective, Lord Peter Fancy, sends his manservant, Bunting, to the Imperial Motorbicycle Hire Company of London W.1, to find out the cost of hiring a motorbike. Bunting subsequently reports: 'I have ascertained, my lord, that motorbicycles may be hired for the day at a cost of 5 guineas, to which must be added one shilling for every mile that the intrepid rider covers'. (Note: in the old British coinage, £1 was 20 shillings and a 'guinea' was 21 shillings.)

(a) Derive a formula that gives the cost in guineas of the hire of a motorbike for a day, in terms of the number of miles ridden.

The plot hinges on whether the suspect could have hired the motorbike to travel from London to Norfolk to commit the crime when he claimed to have been in London at the time. Lord Peter finds the stub of a cheque made out to the Imperial Motorbicycle Hire Company for 12 guineas.

(b) By rearranging the formula from (a), and substituting for y, work out how far the suspect had travelled on his hired motorbike that day.

4.3 Solving equations

The methods for rearranging formulas can also be used in solving equations. This is illustrated by the following problem.

As you may recall from Activity 41, Ann was 25 years old when her son Kim was born. Therefore, to find Kim's age, 25 has to be subtracted from Ann's age. The other day, Ann and her father were turning out some old papers and they came upon a family photograph taken on Kim's eighth birthday, with his first bike. 'You've changed a bit since then, Ann,' observed her father. 'How old were you when that photo was taken?'

Although this problem could be solved easily without algebra, it is useful to take an algebraic approach in order to demonstrate the methods that are needed in more complicated situations.

Ann's age when the photograph was taken is unknown: let x denote her age at that time. So Kim's age when the photograph was taken must have been $x - 25$. But it is known that he was eight years old at that time; therefore

$$x - 25 = 8.$$

This is called an *equation* in x, and it can be *solved* to find how old Ann was when the photo was taken. The value of x that satisfies the equation $x - 25 = 8$ is called the *solution*, and the process of finding that value is called *solving the equation*.

The nineteenth-century mathematics educator Mary Boole described one essence of algebra as 'working with the as-yet-unknown'.

You can think of solving an equation as being the same as making x the subject of the equation and thus obtaining the equation in the form $x =$ some number. It follows that the principle used to solve an equation is the same as the principle used in rearranging formulas: the principle of successively undoing the operations by doing the same to both sides.

In the equation $x - 25 = 8$, the doing operation applied to x is 'subtract 25'. In order to get x on its own, the reverse operation, namely 'add 25', has to be applied to both sides of the equation. This gives

$$x - 25 + 25 = 8 + 25,$$

which simplifies to

$$x = 33.$$

Therefore the solution to the equation is $x = 33$. Hence Ann was 33 years old when the photo was taken.

That may appear to be the end of this particular story, but there is one more thing to mention. There is a simple way of checking that you have got the right answer when solving an equation. When the correct value of x is substituted into the equation, both sides of the equation will give the same number. In this case, 33 is thought to be the solution of the equation $x - 25 = 8$. So if 33 is substituted for x, both sides of the equation should give the same number. Substituting 33 for x on the left-hand side gives $33 - 25$, which is 8 and is indeed equal to the right-hand side. This means there is no mistake about the solution.

In part (b) of Activity 45, you were asked to rearrange a general formula in order to find the value of one variable when the value of another was given. The formula was $y = 5 + \frac{x}{21}$, and the problem was to find the value of x when y takes the value 12. You may have felt that it was a bit unnecessary to find a general formula giving x in terms of y when all that you were asked for was the value of x for a given value of y. In fact, you could have found this value by solving an equation. Example 16 shows how this is done.

Example 16   *Alternative detective work*

The problem in part (b) of Activity 45 could be expressed as:

$$\text{Find the value of } x \text{ when } 12 = 5 + \frac{x}{21}.$$

The problem now is to solve this equation, and find the value(s) of x that makes the equation correct.

Solution

The equation $12 = 5 + \frac{x}{21}$ could be solved by using exactly the same steps as those employed when you changed the subject of the equation $y = 5 + \frac{x}{21}$ in Activity 45, but with y replaced by 12 throughout. However, it is usually easier to simplify the calculations as you go along, as shown below.

Starting with

$$12 = 5 + \frac{x}{21},$$

subtract 5 from both sides and get

$$12 - 5 = 5 + \frac{x}{21} - 5,$$

or

$$7 = \frac{x}{21}.$$

Then multiply both sides by 21:

$$7 \times 21 = \frac{x}{21} \times 21.$$

This gives

$$147 = x.$$

So the solution is $x = 147$. (This is also the value of x found when solving the problem in Activity 45.)

To confirm that this is the correct solution of the equation $12 = 5 + \frac{x}{21}$, substitute 147 for x in the right-hand expression to get $5 + \frac{147}{21} = 5 + 7$. Since $5 + 7 = 12$, the right-hand and left-hand sides of the equation are equal, confirming that 147 is indeed a solution.

Activity 46   *A visit from Yim*

A Singaporean colleague, Yim, was visiting the UK recently. In Singapore, distances are given in kilometres so he does not have a feel for miles.

I invited him to my house for a social visit and he asked how far it was from where he was staying. I told him 12 miles, but he wanted me to tell him the distance in kilometres. The way I remember the conversion factor is this: 8 kilometres are about 5 miles.

(a) If the distance to my house is x kilometres, write down the conversion equation that x must satisfy.

(b) Solve the equation to find out how far, in kilometres, Yim had to travel to my house.

   (Do not forget to check your solution.)

If you are not familiar with solving equations, you might find it helpful to standardize things as far as possible and make sure that the variable is on the left-hand side to begin with. For instance, before solving the equation in Example 16, you could have changed $12 = 5 + \dfrac{x}{21}$ to

$$5 + \frac{x}{21} = 12.$$

Carrying out the undoing operations on this equation would then result in

$$x = 147,$$

with x appearing on the left-hand side in the standard position.

As you get more experienced you may not need to list the doing and undoing steps separately before applying the undoing steps to both sides of an equation.

Activity 47 *An equation to solve*

Solve the equation $120 = 5 + \dfrac{x}{21}$. (Do not forget to check your solution.)

Activity 48 *More equations to solve*

Solve the following equations:

(a) $2x + 1 = 5$

(b) $\frac{1}{2}x - 1 = \frac{1}{2}$

(c) $7x - 3 = 0$

(Do not forget to check each solution.)

Activity 49 *Odd one out*

Which of the following is the odd one out?

$$1 + 3x = 7 \qquad 7 = 3x + 1 \qquad 3x + 7 = 1 \qquad 7 = 1 + 3x$$

Activity 50 *A cold problem*

The conversion formula from Fahrenheit to Celsius is $c = (f - 32) \div 1.8$. Use this formula to write down an equation for the Fahrenheit equivalent of a temperature of $^{-}20\,^{\circ}\mathrm{C}$ (that is, 20 degrees below zero). Solve the equation.

The equations you have met in this unit have all had the form

an expression in x = a number.

However, frequently equations occur in the form

an expression in x = another expression in x,

as the following example illustrates.

Example 17   *Ali's age?*

Ali is 23 years older than his daughter. She is grown up now. A week or so ago, on her birthday, he realized that his age was exactly double hers. He mentioned this, but as he is a bit shy about his age, he did not reveal his actual age. Calculate Ali's daughter's age and thus find how old Ali is.

Solution

Use x to denote the daughter's age in years. Now Ali is 23 years older than his daughter, so his age is $x + 23$ years. In addition, you know that he is twice as old as her, so his age can also be expressed as $2x$ years. There are now two expressions for the same thing—Ali's age. These two expressions must be equal: thus

$$2x = x + 23.$$

This is the equation to use to find x.

Unlike in earlier examples, the 'doing and undoing' method cannot be used here as x occurs in more than one place. To solve the equation, all the x terms need to be gathered together. To do this, use the method of 'doing the same to both sides', applying the operations not just to the numbers but also to the terms involving x.

So subtract x from both sides to obtain

$$2x - x = x + 23 - x,$$

which simplifies to the solution

$$x = 23.$$

Therefore Ali's daughter is 23 and, as Ali's age is twice this, he must be 46.

Check this value against the original problem. Ali's daughter is 23, and Ali is 23 years older and so is 46. This shows that 46 is the solution to the original problem.

Note that it is better to check that the solution fits the original problem, rather than the equation. That way you may spot any mistakes in setting up the equation.

Activity 51

Solve the following equations for x:

(a) $3x = x + 24$.   (b) $5x - 6 = 2x$ (c) $4 + 9x = 18 + 2x$

It is important to appreciate that the same basic rules which were used for rearranging formulas also apply to the solution of more general equations. The art of solving equations by using algebra involves employing the rules of rearrangement to get the equation into the form $x = $ some number. This may take several steps; a sensible intermediate aim is to rearrange the equation so that there are only multiples of x on one side, and only numbers on the other side.

Example 18   *Fahrenheit = Celsius?*

Is there a temperature that has the same numerical value in Fahrenheit as in Celsius?

Solution

Call this unknown numerical value of the temperature x.

Recall that $c = (f - 32) \div 1.8$. When $x\,°\mathrm{F}$ is the same temperature as $x\,°\mathrm{C}$, then both c and f equal x. Therefore x must satisfy the equation

$$x = (x - 32) \div 1.8.$$

The task is to solve this equation. All the multiples of x need to be on one side of the equation, and all the numbers on the other. Collect the multiples of x together on the left-hand side. However, the x term on the right-hand side of the equation is bound up inside brackets and, before it can be moved, it must be 'set free'. So the very first step is to multiply through by 1.8 to get rid of the division on the right-hand side of the equation. This gives

$$1.8x = x - 32.$$

To get all the terms involving x on the left, x is subtracted from both sides, giving

$$1.8x - x = {}^{-}32.$$

Now x is the same as $1x$, so $1.8x - x = 0.8x$. Therefore the equation becomes

$$0.8x = {}^{-}32.$$

To get x on its own, undo the multiplication by 0.8. This involves dividing both sides by 0.8, to obtain

$$x = {}^{-}32 \div 0.8,$$

or

$$x = {}^-40.$$

The algebra has shown that $^-40$ is the solution of the equation.

Thus $^-40\,°\text{F}$ is the same as $^-40\,°\text{C}$. (There are no other numbers like this, as there are no other solutions to the equation.)

> There are often several routes for solving an equation, but regardless of how you carry out the process in detail, the following overall strategy is useful:
> - simplify the equation to get rid of fractions, division and brackets;
> - get the x terms together on one side of the equation, and the numbers together on the other side;
> - collect like terms;
> - divide both sides by the coefficient of x.

Remember that whenever you solve an equation, you should check the solution. The test that a solution has to meet is this: when the value of the solution is substituted for x wherever x appears in the equation, the resulting numerical values of the expressions on either side of the equation must be equal.

The three activities below (Activities 52–54) provide exercises for you to practise checking the solutions of equations, solving equations and formulating equations. The subsequent activities (Activities 55–57) then combine these techniques.

As you are working on these activities, think about the strategies that you are using, and why they apply. Do not forget to check your answers by substituting the solution into the original problem (*before* turning to the solutions at the end of the unit). It is not enough just to read about algebra. If you are to learn how to use it, you must practise the skills for yourself.

As you practise, add information about the techniques to your Handbook notes.

Activity 52   *Checking solutions*

(a) Confirm that $x = {}^-40$ is a solution of $x = (x - 32) \div 1.8$.

(b) Verify that 40 is *not* a solution.

Activity 53   *Solving equations*

Solve the following equations:

(a) $3x = x + 4$ (b) $x + 1 = 2x - 1$

(c) $3(x + 1) = 2x - 1$ (d) $5x \div 4 = x + 3$

(e) $7y + 1 = 17 - y$ (f) $y \div 2 + 3 = y - 1$

(Check your solutions.)

Here is an example that involves all of the techniques.

Example 19   *Mark has triplets*

Mark has triplets. They were born on his 22nd birthday. However, on their most recent joint birthday, he realized that his age was equal to the sum of their ages.

(a) Formulate an equation for the age, x, of the triplets.

(b) Solve this equation.

(c) Check that the solution satisfies the general problem.

Solution

(a) If the triplets are each x years old, then Mark will be $x + 22$ years old. But the sum of the triplets ages is $x + x + x$, or $3x$. This sum is equal to Mark's age. So the required equation is $3x = x + 22$.

(b) To solve this equation, get all the terms in x on one side by subtracting x from both sides:

$$3x - x = x + 22 - x.$$

This gives

$$2x = 22.$$

So

$$x = 11.$$

(c) If the triplets are 11 years old, Mark, whose age can be expressed as $x + 22$, will be $11 + 22$, or 33 years old. Now, Mark's age is also equal to sum of the triplets ages, that is $11 + 11 + 11 = 33$. Therefore the solution is correct.

Activity 54   *Formulating and checking equations*

(a) Janet is ten years older than Leroy. In two years' time, she will be twice as old as Leroy. Formulate an equation that will enable you to work out how old Janet is now.

Check that Janet's age is eighteen.

(b) In my house there are as many (four-legged single-tailed) pets as there are (two-legged) people. If I multiply the number of heads by the number of tails, I get the number of feet. Find an equation for the number of people, x, (or, equivalently, the numbers of pets) in the house.

Check that I have three pets.

Activity 55   *Formulating and solving equations*

(a) At 10 a.m., Alastair is 75 miles from Cardiff, heading towards Edinburgh at 50 mph, while Bethan is in Edinburgh, 400 miles from Cardiff, poised to start driving towards Cardiff at 60 mph on the same route. As you may remember, this means that t hours after 10 a.m., Alastair's distance from Cardiff (in miles) will be $75 + 50t$, while Bethan's distance from Cardiff will be $400 - 60t$. Write down an equation that gives the time at which they will meet.

Recall Section 1.1, p.27.

(b) Solve the equation to predict when Alastair and Bethan will rendezvous, correct to the nearest quarter of an hour.

Activity 56   *Fahrenheit equals Kelvin*

Recall that a conversion formula from Kelvin to Fahrenheit is $f = 1.8(k - 273) + 32$, where a temperature of f °F is k on the Kelvin scale.

Is there a temperature, x, that has the same numerical value on both scales?

Use a similar method to the one in Example 18 to solve the equation for x.

[*Hint*: first multiply out the expression involving brackets, $1.8(k - 273)$. You may need your calculator for the arithmetic.]

Activity 57   *A Shakespearean problem*

Omit this activity if you are short of time.

Shakespeare memorably described the seven ages of man in *As You Like It*, act II, scene vii.

Suppose that a man spends a fifteenth of his life as an infant 'mewling and puking in the nurse's arms'; two-fifteenths of his life as a schoolboy 'creeping like a snail unwillingly to school'; a twenty-fifth as a lover; a fifth as a soldier; two-fifths as a justice; ten years as a 'lean and slipper'd pantaloon'; and that the final two years of his life are spent in 'second childishness and mere oblivion, *sans* teeth, *sans* eyes, *sans* taste, *sans* everything'. What is the life expectancy of a man?

Check your solution by finding the length of each of the seven ages in years.

Activity 58   *Reviewing progress*

Before moving on, review your progress in algebra. Check your Handbook notes and add any further points. How confident do you feel about using the techniques you have just been practising? Has any aspect been particularly difficult? What feedback have you used, and how did it help? (For example, you may have been working with other students or your tutor, or used the comments in the unit itself.) Do you need more practice? If so, look in *Resource Book B* for suitable exercises.

Outcomes

After studying this section, you should be able to:

◇   undo operations taking account of the order (Activities 39–44);

◇   change the subject of a formula (Activities 39–44);

◇   solve an equation algebraically in simple cases, and check the solution (Activities 45–54);

◇   formulate an equation (Activities 45, 46, 54, 55 and 57).

5 Algebra and your calculator: graphs

Aims   This section aims to teach you how to use your course calculator to produce graphs that represent functions, and how to use these graphs to solve equations.   ◇

5.1 Using the calculator to draw graphs

From Section 3, you should know how to enter and store a function (or formula) in your calculator, as well as how to use the calculator to evaluate the function for any chosen numerical value of the variable, and how to draw up a table evaluating the function for a number of different values of the variable. The next step is to get the calculator to draw a graph of the function. Following that, you will be shown how to use the calculator to explore the features of a graph and, in particular, how to use the graphical facilities of the calculator to solve equations.

In *Unit 7*, a function was called 'a mathematical relationship'.

The process of drawing a graph to represent a function was introduced in *Unit 7*. For each value of x, the corresponding value of y is calculated using the formula that specifies the function, and then the point whose coordinates are (x, y) is plotted. The collection of points plotted in this way forms the graph. In many of the cases considered in this section, the result is a curve, not a straight line.

When drawing a graph by hand on graph paper, you can plot only a limited number of points, and to obtain the graph you have to join up these points smoothly. However, with the calculator, there is the advantage that many more points are plotted, giving the appearance of a continuous curve.

 Work through Sections 8.4, 8.5 and 8.6 of Chapter 8 of the Calculator Book.

5.2 Zooming around the course calculator

Altogether, four new calculator skills have been covered in this unit: *entering and storing functions*; *creating tables of values*; *drawing graphs*; and *solving equations graphically* (using the trace and zoom facilities). Individually, these are important skills, but in many situations you need to use most or all of them one after the other. This can be rather demanding.

The next activity involves using all of these skills. To help you with it, there is an audio sequence. You will be asked to enter, tabulate and produce a graph of quite a complicated function, and to solve an equation by finding where the function takes the value zero. The speaker will talk you through the whole activity. In addition to the audio CD and player, you will need only your calculator, though you might find it useful to have a piece of paper and a pen or pencil to hand. For convenience, jot down

the formula for the function, so that you do not have to keep referring
back to the unit.

Activity 59   *Trace and zoom*

Use band 3 of CDA5509 (Track 15) to help you plot the function

$$y = 1 - \frac{3x}{x^2 + 1}.$$

Use your graph to solve the equation

$$1 - \frac{3x}{x^2 + 1} = 0.$$

5.3   Things to do

Activity 60   *Conversion tables*

If a distance is M miles or, equivalently, K km, then M and K are related
by the formula

$$K = 1.61M.$$

Use your calculator to produce tables giving the kilometre equivalents of
distances in miles, and also produce conversion graphs suitable for the
following purposes:

(a) use by someone familiar with the metric system but not the imperial
system, who is touring the UK by train, bus or car;

(b) use by a parish or town council for converting the distances given on
signposts for footpaths and cycle tracks into kilometres.

You have now seen three ways of representing the process of converting
distances in miles to the corresponding measurements in kilometres: by
means of the formula $K = 1.61M$ (see Section 1.2); by means of a table
(see Activity 60); and by means of a conversion graph (first encountered in
Unit 7). This is a good point to pause and compare these different
methods.

Each of the three representations of the conversion (a symbolic formula, a
numerical table and a graph) is useful in its own way, but each has its
advantages and disadvantages according to the context and the purpose for
which it is being used.

The symbolic formula is very accurate, in principle: the only limits on its
accuracy derive from the accuracy with which the conversion factor (1.61

in the present example) is known, and the accuracy with which calculations can be carried out. Furthermore, there is no limit, in principle, to the sizes of the distances that can be converted: the formula can as easily be used for very small distances—two or three miles—as for very large distances—thousands of miles, or more. However, using a formula can be time-consuming, especially if there are a huge number of distances to convert.

A table is quick to use, but it can only be employed directly to convert those distances that actually appear in it. If you want approximate answers—suppose that distances to the nearest ten miles are good enough—then a table is very useful.

Like a table, a graph also covers only a specific range of distances. However, a graph can be used to convert intermediate distances that do not appear in the table. Also you can convert distances either from miles to kilometres or from kilometres to miles using the same graph. The graph has the added advantage of showing pictorially how miles and kilometres are related. Although a formula also reveals the nature of this relationship, you have to be able to understand formulas before you can appreciate them in this way.

Activity 61   Cube root

In this activity, you will use your calculator to find the *cube root* of 5. The cube root of 5, written $\sqrt[3]{5}$, is the number whose cube is 5: in other words, it is the solution of the equation $x^3 = 5$. The method used here is to plot $y = x^3$ and $y = 5$ simultaneously, and find the x-coordinate of the point where the graphs intersect.

Store the functions $y = x^3$ and $y = 5$ in your calculator, and plot them both. You will find that the graph of $y = 5$ is a horizontal line, which crosses the curve representing $y = x^3$ somewhere between $x = 1.6$ and $x = 1.8$. Find the x-coordinate of the point of intersection, correct to 2 decimal places.

Check your answer by calculating its cube. Alternatively, calculate $\sqrt[3]{5}$ directly on your calculator.

Activity 62   *Drug doses for children*

In order to calculate the dose, for a child, of a particular medication, the following procedure is sometimes used:

- weigh the patient in kilograms;

- divide this by 70 (the average adult weighs 70 kg);

- take the cube root of the result;

- square the result;

- multiply by the standard adult dose.

This gives the dose for a child weighing a specific amount.

(a) If the standard daily dose of phenobarbital for the average adult with a certain type of epileptic illness is 100 milligrams, write down a formula for the dose d (in milligrams) for a child weighing w kg. (Use the steps listed above as though you were working through a 'think of a number' trick, with the initial number being w.)

(b) Plot this function on your calculator for w with values from 0 to 70. Hence estimate the daily dose for a child weighing 35 kg.

(c) Theophylline is a drug that is sometimes used to treat asthma. In a particular clinical situation, the maintenance dose for an average adult is 200 milligram every 4 hours. Amend your formula from part (a) to obtain a formula for the dose of theophylline for a child weighing w kg. Use your calculator to produce a table of doses for children of different weights, to aid a hospital department.

Outcomes

After studying this section, you should be able to:

◇   use your calculator to produce graphs and tables of functions over a given range of the independent variable (Activities 60–62);

◇   use your calculator to solve equations graphically (Activity 60).

Unit outcomes

Activity 63 *Looking back*

Before you leave this unit, check whether you have anything to complete. Look again at the activity sheet on the pros and cons of algebra, and at your notes on terms and techniques; make sure they are all up to date.

It is often useful to take time to think about what you have been doing and what you have learned, before you rush on to the next piece of work. Run through the unit in your mind's eye and make a mental list of what it covered, including the new vocabulary and techniques.

Look again at your activity sheet for Activity 1. How have you progressed in algebra? How could you demonstrate your progress? Think about which responses to activities you could draw on for this purpose. If you have been using the outcomes to help monitor your progress, which ones are you confident about and which (if any) do you still need to master? How do you feel about algebra now?

Outcomes

After studying this unit, you should be able to:

◇ build up formulas from given information, including from the steps of a 'think of a number' sequence;

◇ recognize whether or not a collection of symbols is a valid algebraic expression;

◇ understand the terms 'substitute', 'evaluate', 'value', 'mathematical function' and 'variable';

◇ evaluate an algebraic expression by substituting a value for the variable;

◇ simplify algebraic expressions, multiply out brackets, and multiply or divide an algebraic expression by a number;

◇ change the subject of a formula;

◇ formulate an equation;

◇ enter and store formulas in your calculator;

◇ use your calculator to evaluate a stored formula for particular values of the independent variable;

◇ use your calculator to produce graphs and tables of functions;

◇   solve equations algebraically or graphically;

◇   use and interpret formulas, tables and graphs in a variety of
     contexts, such as converting units, travel planning, calculating
     areas and volumes, and adjusting drug dosages for body mass;

◇   make notes about various aspects of learning algebra.

Comments on Activities

Activity 1

It is often useful to think about what you already know in relation to an area of work so that you can use it as a bridge into the new work. Sometimes it is also helpful to consider how you feel about a particular area. Some students have strong feelings about algebra, and this may create a 'block'. By thinking about and analysing what causes a difficulty or block, it is often overcome.

You may wish to use your activity sheet as a log of your progress in algebra during your study of this unit. You could cite completed activities as evidence of your continuing progress.

Activity 2

What you write will depend very much on your previous experience, but you may have included the following points:

- multiplication and addition are commutative, so, for example, $2 \times 3 = 3 \times 2$ and $4 + 3 = 3 + 4$;
- multiplication and division are distributive over addition, so multiplying out brackets gives $3 \times (2 + 4) = 3 \times 2 + 3 \times 4$;
- letting a letter stand for a number in a 'think of a number' trick makes it easy to analyse why the trick works.

Activity 3

(a) Assume a volume measures G gallons, or L litres. Since 1 gallon is 4.55 litres, G gallons are $4.55 \times G$ litres. But G gallons are the same quantity as L litres. Hence

$$L = 4.55 \times G.$$

(b) Take G to be $\frac{1}{8}$. Then, using the formula obtained in part (a),

$$L = 4.55 \times \frac{1}{8}$$
$$= 0.57 \text{ (to 2 d.p.)}.$$

So 1 pint is about 0.57 litre.

Activity 4

Since 1 gallon is 4.55 litres (see Activity 3), 1 litre is $1 \div 4.55$ gallons. So L litres are $L \div 4.55$ gallons.

But L litres are the same quantity as G gallons. Hence

$$G = L \div 4.55.$$

Now, in Activity 3(b), $L = 0.57$ litre, and so

$$G = 0.57 \div 4.55$$
$$= 0.125 \text{ (to 3 d.p)}.$$

Since 0.125 is the same as $\frac{1}{8}$, this confirms the answer to Activity 3(b).

Activity 5

If you put $C = 6\frac{1}{2}$ in the formula $M = 0.46 \times C$, then $M = 2.99$. So Goliath was about 3 m tall!

The explanation might be something like: Goliath was six and a half cubits tall (a span is half a cubit). One cubit is about 0.46 m. So multiply the number of cubits in Goliath's height by 0.46 to get his height in metres:

$$0.46 \times 6\frac{1}{2} = 2.99.$$

As you can see, the formula is much more concise than the word explanation.

Activity 6

(a) A temperature of $98.4\,°\mathrm{F}$ means $f = 98.4$.

Substituting for f in the given formula produces

$$c = (98.4 - 32) \div 1.8$$
$$= 66.4 \div 1.8$$
$$\simeq 36.9 \text{ (to 1 d.p.)}.$$

So $98.4\,°\mathrm{F}$ is about $36.9\,°\mathrm{C}$.

(b) A temperature of $50\,°\mathrm{C}$ means $c = 50$. Substituting for c in the given formula produces

$$f = 50 \times 1.8 + 32$$
$$= 90 + 32$$
$$= 122.$$

So $50\,°\mathrm{C}$ is $122\,°\mathrm{F}$.

To check that this is correct, put $f = 122$ in the formula. This gives

$$c = (122 - 32) \div 1.8$$
$$= 90 \div 1.8$$
$$= 50,$$

as expected.

Activity 7

Include the idea of a mathematical function (input, process or rule, and output) and the use of a formula to represent such a function. You might use examples from the conversion formulas in Examples 1, 2 and 3 and/or Activities 3, 4, 5 and 6. Diagrams like Figures 1–3 and 5 might also be useful.

Activity 8

(a) A suitable formula for converting from K kilometres per hour to M miles per hour is $M = K \div 1.61$. With $K = 300$, it follows that

$$M = 300 \div 1.61$$
$$\simeq 186.$$

So $300\,\mathrm{km/h}$ is about $186\,\mathrm{mph}$.

(b) Let a speed of $K\,\mathrm{km/h}$ be equivalent to $N\,\mathrm{m/s}$. If you travel K km in an hour, then you travel $1000 \times K$ metres in an hour. As there are $60 \times 60 = 3600$ seconds in an hour, your speed (in m/s) is $1000 \times K \div 3600$. So

$$N = 1000 \times K \div 3600$$
$$= 0.28 \times K \text{ (to 2 s.f.)}.$$

(An alternative way of writing the formula is $N = K \div 3.6$.)

(c) A speed of $300\,\mathrm{km/h}$ means $K = 300$. So

$$N = 0.28 \times K$$
$$= 0.28 \times 300$$
$$= 84.$$

Therefore a speed of $300\,\mathrm{km/h}$ is equivalent to about $84\,\mathrm{m/s}$.

(Eurostar travels $84\,\mathrm{m}$ every second!)

Activity 9

(a) Time taken = distance travelled ÷ average speed.

The given average speed is $s\,\mathrm{km/h}$. When the distance travelled is $1\,\mathrm{km}$, then

$$\text{time taken } = 1 \div s \text{ hours}.$$

There are 3600 seconds in an hour, so time taken in seconds, t, is given by

$$t = 3600 \div s,$$

or

$$t = \frac{3600}{s}.$$

(b) (i) For this Eurostar train, $s = 300$. So the train takes $3600/300 = 12\,\mathrm{s}$ to cover a kilometre.

(ii) For the Earth, $s \simeq 107\,000$. So the Earth takes $3600 \div 107\,000 \simeq 0.034\,\mathrm{s}$ to travel a kilometre.

(iii) For light, $s \simeq 1.1 \times 10^9$. So light takes $3600 \div (1.1 \times 10^9) \simeq 0.000\,0033\,\mathrm{s}$ (to 2 s.f.) to travel a kilometre.

Activity 10

The time interval from 10 a.m. to 2.15 p.m. is $4\frac{1}{4}$ or 4.25 hours.

Alastair's distance, a miles, from Cardiff at 2.15 p.m. is given by

$$a = 75 + (50 \times t)$$
$$= 75 + (50 \times 4.25)$$
$$= 287.5.$$

Therefore, Alastair will be 287.5 miles from Cardiff.

Bethan's distance, b miles, from Cardiff is given by

$$b = 400 - (60 \times t)$$
$$= 400 - (60 \times 4.25)$$
$$= 145.$$

Therefore, Bethan will be 145 miles from Cardiff.

Hence, at 2.15 p.m., Alastair will be further from Cardiff (and nearer to Edinburgh) than Bethan.

Activity 11

(a) When the time is t hours, you have been travelling for $(t - 10)$ hours. When the trip meter reads d miles, you have travelled a distance of $(d - 100)$ miles. Your average speed, s mph, is given by

$$s = \frac{(d - 100)}{(t - 10)}, \quad \text{or} \quad s = \frac{d - 100}{t - 10}.$$

(b) At a quarter past eleven, the time t in decimals (*not* in hours and minutes) is 11.25. From the trip meter reading, $d = 180$. So the average speed since 10.00 (in mph) is given by

$$s = \frac{(180 - 100)}{(11.25 - 10)} = \frac{80}{1.25} = 64.$$

So your average speed since 10.00 is 64 mph.

Activity 12

Your plan of the lawn should look something like this:

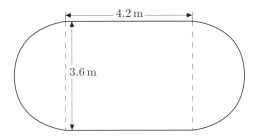

The area of the rectangular piece of lawn is

$$\text{length} \times \text{breadth} = 4.2 \times 3.6 \, \text{m}^2 = 15.12 \, \text{m}^2.$$

The two semi-circular pieces combine to form a circle of diameter 3.6 m, with

$$\text{area} = \pi \times (\text{radius})^2$$
$$= \pi \times (1.8)^2 \, \text{m}^2 = 10.18 \, \text{m}^2.$$

So the total area of the lawn is $25.3 \, \text{m}^2$, but you should probably round it up to $26 \, \text{m}^2$ when ordering the turf.

Activity 13

(a) In symbols the formula is

$$V = \frac{4}{3} \times \pi \times r^3.$$

However, the multiplication signs can be omitted and the formula can then be written as

$$V = \frac{4}{3} \pi r^3.$$

(b) The two circular ends, each of radius r, have a combined area of $2 \times \pi \times r^2$. The curved surface when laid out flat becomes a rectangle with sides l and $2\pi \times r$ (one side is the circumference of the circle).

Therefore the total surface area, A, of the cylinder is given by

$$A = (2 \times \pi \times r^2) + (l \times 2\pi \times r).$$

If you omit the multiplication signs, this becomes

$$A = 2\pi r^2 + 2\pi rl.$$

(c) The area of the base of the cylinder is $\pi \times r^2$. So the volume, V, of the cylinder is given by

$$V = \pi \times r^2 \times l$$

or, if you omit the multiplication signs, by

$$V = \pi r^2 l.$$

Activity 14

(a) The area of the curved surface is $2\pi \times r \times l$, and the area of the top, which is a circle, is $\pi \times r^2$. So the area of the curved surface for the given dimensions is

$$2\pi \times 0.25 \times 1 \simeq 1.6 \, \text{m}^2,$$

and the area of the circular top is

$$\pi \times 0.25^2 \simeq 0.20 \, \text{m}^2.$$

Thus the total area is $1.8 \, \text{m}^2$, which means that the 1 m square and 1 m by 1.5 m pieces of card are too small.

To decide which card Niels should buy, you need to consider the dimensions of the different parts of the cardboard cylinder, and think how these would be best accommodated on the various sizes of card.

The dimensions of the curved surface of the cylinder are 1 m by $2\pi \times 0.25$ m, which is 1 m by about 1.6 m. The longest dimension is 1.6 m, so a card that is 1 m by 2 m or 1.5 m by 2 m should be bought.

The circular top has a diameter of $2 \times 0.25 \, \text{m} = 0.5 \, \text{m}$; it could therefore be cut out of a square piece of card of side 0.5 m.

To get both pieces out of a single sheet of card, the 1.5 m by 2 m card should be bought. However, there will be a lot of wastage with this size of card, so reducing the radius of the cylinder slightly and thus making the longest dimension no more than 1.5 m would enable the 1.5 m square size of card to be used.

(b) The height of the can, 0.8 m, can be accommodated along the 1 m dimension of the card. So the top of the can could have a circumference of 1.1 m along the other dimension of the card. If r is the radius of the can, then $2\pi \times r = 1.1$ m. Hence $r = 1.1 \div (2\pi) \simeq 0.18$ m.

Therefore Sujatha could make a can of radius 0.18 m.

Activity 15

(a) The formula for Hooper's rule is

$$A = 110 \times N + 30.$$

(b) With 3.5 species, on average, per 30 yards of hedge, $N = 3.5$. From the formula for Hooper's rule, the age of the hedge is therefore about $110 \times 3.5 + 30 = 415$ years. In other words, it dates from just before the year 1600 according to Hooper's rule.

Activity 16

Here are some of the advantages and drawbacks of algebra. You may have thought of others.

Advantages

- It is quicker and more concise than writing word formulas.
- You can see relationships more clearly.
- It is easy to apply an algebraic formula by substituting number for letters.
- It is much easier to show why number tricks work, than by using words alone.

Drawbacks

- You may forget what the letters stand for. (So try to write down what each letter denotes.)
- The rules for manipulating algebra are not as familiar as those in ordinary language.

Activity 17

(a) When $N = {}^{-}1$, then
$$4 \times N + 10 = 4 \times {}^{-}1 + 10 = {}^{-}4 + 10 = 6.$$

(b) When $x = 6$, then
$$1 + \frac{1}{3} \times x = 1 + 2 = 3.$$

(c) When $y = 3$, then
$$12 \div y + 4 = 4 + 4 = 8.$$

Activity 18

(a) When $M = 3$, then
$$3 \times M + 5 = 3 \times 3 + 5 = 14.$$

So the value of the expression is 14.

(b) (i) When p has the value 3, then $2 \times p - 2$ has the value $2 \times 3 - 2$, which is $6 - 2$, or 4. (Be careful to multiply by 2 before subtracting 2.)

(ii) When p has the value 7.165, the expression has the value 12.33.

(iii) When p has the value $^{-}1$, the expression has the value $^{-}4$.

(c) When b is 5, then
$$2 + b - 2 \times b = 2 + 5 - 2 \times 5 = {}^{-}3.$$

So the value of the expression is $^{-}3$.

(d) When $a = 3$ and $b = 5$, then
$$a \times b - a - b + 1 = 15 - 3 - 5 + 1 = 8.$$

So the value of the expression is 8.

Activity 19

(a) This does not make sense: there is a 'floating' minus sign at the end.

(b) This makes sense.

(c) This makes sense: the $+$ is fairly pointless, but not actually wrong (you might want to emphasize that the first number is $+4$ and not $^{-}4$).

(d) This makes sense: the $^{-}$ indicates a negative number.

(e) This does not make sense: there needs to be a number between $\times$ and $\div$.

(f) This does not make sense unless one of the $\div$ signs is deleted.

(g) This does not make sense: there needs to be a number between the $-$ and $\times$ signs.

Activity 20

(a) $3N$ (b) $\frac{4}{6}z$, or $\frac{2}{3}z$ (c) $^{-}4N$

(d) $48x$ (e) $^{-}10N$ (f) $^{-}2x$

(g) $56P^2$ (h) $10x^3$

Activity 21

There is no comment on this activity. See the discussion in the main text.

Activity 22

(a) $24A$ (b) $21B$ (c) $10x$ (d) $6y$

(e) $7p^2$ (f) $6A + 4B$ (g) $8A + 4B$

(h) $6A + 7A^2$

Note that the expressions (f) and (h) do not simplify.

Activity 23

(a) Here $x + 8x$ gives $9x$, and $3x^2 - 2x^2$ gives x^2. So the expression simplifies to
$$9x + x^2.$$

(b) Collecting like terms x and $3x$, and $8x^2$ and $^{-}2x^2$, gives
$$4x + 6x^2.$$

(c) Collecting like terms 1 and 5, and $2a$ and $4a$, gives
$$6 + 6a + 3a^2.$$

(d) Multiplying out the first and last terms gives $6A + 2A^3 + 8A$, which is

$$6A + 8A + 2A^3, \quad \text{or} \quad 14A + 2A^3.$$

(e) Multiplying out the terms gives

$$10A^2 - 6A^2 = 4A^2.$$

(f) Multiplying out the first and last terms gives
$$^-3x^2 + x - 8x^2 = x - 3x^2 - 8x^2$$
$$= x - 11x^2.$$

(g) Take the square root of the last term and then collect like terms:

$$2z + x^2 + z = 3z + x^2.$$

(h) Take the square root of the fourth term and then collect like terms:

$$y^3 + 2y + y^2 - 2y - 4$$
$$= y^3 + y^2 + 2y - 2y - 4$$
$$= y^3 + y^2 - 4.$$

(It does not matter if you have the terms in a different order: for example, $3a^2 + 6a + 6$ is the same as $6 + 6a + 3a^2$.)

Activity 24

$4p + 5 + p - 3 = 5p + 2$.

When $p = 5$, $4p + 5 + p - 3 = 27$, and $5p + 2 = 27$.

Activity 25

(a) When $N = 7$,

$$4(N + 10) = 4 \times 17 = 68,$$

and

$$4N + 10 = 4 \times 7 + 10 = 38,$$

and

$$N + 10 \times 4 = 7 + 40 = 47.$$

(b) 'Double $2N + 5$' can be written as

$$2(2N + 5).$$

'Divide $4N + 4$ by 4' can be written as

$$(4N + 4) \div 4,$$

or

$$(4N + 4)/4,$$

or

$$\frac{(4N + 4)}{4}.$$

Activity 26

(a) (i) $9(m + 3)$ (ii) $(m + 3)(9 + 2m)$

(b) In all cases, collect like terms.

  (i) $5(m + 3)$ (ii) $3(4N + 10)$

  (iii) $3(9 - 2z) + 2(9 - 2z) = 5(9 - 2z)$

Activity 27

(a) $2m + 6$ (b) $12X + 6$ (c) $6x - 12$

(d) $n^2 + 2n$ (e) $12X^2 + 6X$ (f) $45 + 10z$

(g) $6m - 18$ (h) $5p - 35p^2$

Activity 28

(a) $(y + 7)(y + 1) = y^2 + y + 7y + 7$
$$= y^2 + 8y + 7$$

(b) $(p + 2)(p + 3) = p^2 + 3p + 2p + 6$
$$= p^2 + 5p + 6$$

(c) $(2x + 7)(x + 1) = 2x^2 + 2x + 7x + 7$
$$= 2x^2 + 9x + 7$$

(d) $(p + 2)^2 = p^2 + 2p + 2p + 4$
$$= p^2 + 4p + 4$$

(e) $(4p + 2)(3p + 2) = 12p^2 + 8p + 6p + 4$
$$= 12p^2 + 14p + 4$$

(f) $(a + b)(c + d) = ac + ad + bc + bd$

(g) $(2a + b)(c + 3d) = 2ac + 6ad + bc + 3bd$

(h) $(3x + 4y)(2w + 5z) =$
$$6xw + 15xz + 8yw + 20yz$$

Activity 29

(a) $(2x + 3)(x - 1)$

$$= 2x \times x + 2x \times (^-1) + 3 \times x + 3 \times (^-1)$$
$$= 2x^2 - 2x + 3x - 3$$
$$= 2x^2 + x - 3$$

(b) $(x - 1)(4x + 2)$

$$= x \times 4x + x \times 2 + (^-1) \times 4x + (^-1) \times 2$$
$$= 4x^2 + 2x - 4x - 2$$
$$= 4x^2 - 2x - 2$$

From (c) on, the first step in multiplying out the brackets is no longer written out in full as you are probably familiar with the technique by now.

(c) $(2x - 1)(3x + 1)$

$$= 6x^2 + 2x - 3x - 1$$
$$= 6x^2 - x - 1$$

(d) $(x - 1)(x + 2)$

$$= x^2 + 2x - x - 2$$
$$= x^2 + x - 2$$

(e) $(2y + 2)(2y - 2)$

$$= 4y^2 - 4y + 4y - 4$$
$$= 4y^2 - 4$$

(f) $(3z + 4)(3z - 4)$

$$= 9z^2 - 12z + 12z - 16$$
$$= 9z^2 - 16$$

(g) $(a + b)(a - b)$

$$= a^2 + ab - ab - b^2$$
$$= a^2 - b^2$$

Note that (e), (f) and (g) all have the same form, often called 'the difference of two squares'. They can all be written in the form $a^2 - b^2$. In (e), this is $(2y)^2 - 2^2$, and in (f), $(3z)^2 - 4^2$.

(h) $(c - d)(2c + d)$

$$= 2c^2 + cd - 2cd - d^2$$
$$= 2c^2 - cd - d^2$$

(i) $(3 - x)(2 + x)$

$$= 6 + 3x - 2x - x^2$$
$$= 6 + x - x^2.$$

Activity 30

(a) $m^2 + 8 + 15m$ is the odd one out. The others are all different versions of $(m + 3)(m + 5)$, which expands to $m^2 + 8m + 15$.

(b) $3x(x + 2) + 2(3x + 1)$ is the odd one out. The others are all different versions of $(3x + 1)(x + 2)$, which expands to $3x^2 + 7x + 2$.

(c) $y^2 + y - 2$ is the odd one out. The others are all different versions of $(y + 1)(y - 2)$, which expands to $y^2 - y - 2$.

Activity 31

Two straightforward ways of writing the expression are

$$x^2 + 2x + 6 + 3x$$

and

$$x^2 + 5x + 6,$$

as well as all the other possible arrangements of these terms. Further possibilities are $x(x + 3) + 2(x + 3)$, $x(x + 2) + 3(x + 2)$, and $(x + 2)(x + 3)$. You may have also found some other forms of the expression.

Activity 32

(a) $(2x - 3)(x - 1)$

$$= 2x \times x + 2x \times {}^-1 + {}^-3 \times x + {}^-3 \times {}^-1$$
$$= 2x^2 - 2x - 3x + 3$$
$$= 2x^2 - 5x + 3$$

(b) $(3x - 1)(4x - 2)$

$$= 3x \times 4x + 3x \times {}^-2 + {}^-1 \times 4x + {}^-1 \times {}^-2$$
$$= 12x^2 - 6x - 4x + 2$$
$$= 12x^2 - 10x + 2$$

(c) $(1 - x)(4x - 2)$

$$= 1 \times 4x + 1 \times {}^-2 + {}^-x \times 4x + {}^-x \times {}^-2$$
$$= 4x - 2 - 4x^2 + 2x$$
$$= 6x - 2 - 4x^2, \quad \text{or} \quad {}^-4x^2 + 6x - 2,$$
$$\text{or} \quad {}^-2 - 4x^2 + 6x$$

(d) $(x-1)^2$

$$= (x-1)(x-1)$$
$$= x \times x + x \times {}^-1 + {}^-1 \times x + {}^-1 \times {}^-1$$
$$= x^2 - x - x + 1$$
$$= x^2 - 2x + 1$$

(e) $(x-1)(x+1)$

$$= x^2 + x - x - 1$$
$$= x^2 - 1$$

(This expression is the difference of two squares: x^2 and 1^2.)

Activity 33

(a) $\quad (4N+8) \div 4 = 4N \div 4 + 8 \div 4$
$$= N + 2$$

(b) $\quad (6m+3) \div 3 = 6m \div 3 + 3 \div 3$
$$= 2m + 1$$

(c) $\quad 5(4-2z) \div 2 = 5(4 \div 2 - 2z \div 2)$
$$= 5(2 - z)$$
$$= 10 - 5z$$

(d) $\quad (6k-12) \div {}^-3 = 6k \div {}^-3 - 12 \div {}^-3$
$$= {}^-2k + {}^-12 \div {}^-3$$
$$= {}^-2k + 4, \quad \text{or} \quad 4 - 2k$$

Activity 34

(a) Correct.

(b) Incorrect: $6 \times (2N+1) = 12N + 6$.

(c) Correct.

(d) Correct.

(e) Incorrect: either
$N + 3(N+3) = N + 3N + 9 = 4N + 9$, or
$(N+3)(N+3) = N^2 + 6N + 9$.

(f) Incorrect: add a bracket
$(X+2)(X+3) = X^2 + 5X + 6$.

(g) Incorrect: add a bracket
$(X+2)(X+4) = X^2 + 6X + 8$.

(h) Incorrect: it should be
$(X-2)(2X+2) = 2X^2 - 2X - 4$.

(i) Correct.

(j) Incorrect: add a bracket
$(4X-2) \div 2 = 2X - 1$.

(k) Incorrect: add a bracket
$(3X+6) \div 3 = X + 2$.

Activity 35

(a) When $X = 2$, $4X - 7 = 8 - 7 = 1$ and
$2X - 3 = 4 - 3 = 1$. So

$$(4X-7)/(2X-3) = 1 \div 1 = 1.$$

When $X = 1$, $4X - 7 = 4 - 7 = {}^-3$ and
$2X - 3 = 2 - 3 = {}^-1$. So

$$(4X-7)/(2X-3) = {}^-3 \div {}^-1 = +3.$$

Note that a negative number divided by a negative number gives a positive number.

(b) When $X = 7$,

$$\frac{6X-2}{3X+1} = \frac{42-2}{21+1} = \frac{40}{22} = \frac{20}{11}.$$

Activity 36

(a) Represent the positive number you first thought of by N.

Step 1: $\quad N - 1$

Step 2: $\quad (N-1)^2 = (N-1)(N-1)$
$$\qquad\qquad\quad = N^2 - N - N + 1$$
$$\qquad\qquad\quad = N^2 - 2N + 1$$

Step 3: $\quad N^2 - 2N + 1 + 2N = N^2 + 1$

Step 4: $\quad N^2 + 1 - 1 = N^2$

Step 5: $\quad \sqrt{N^2} = N$

Step 6: $\quad N - N = 0$.

If you did not simplify step by step, then the result of following through the steps to the final step is

$$\sqrt{(N-1)^2 + 2N - 1} - N.$$

Now

$$(N-1)^2 = N^2 - 2N + 1,$$

so

$$(N-1)^2 + 2N - 1 = N^2,$$

and therefore

$$\sqrt{(N-1)^2 + 2N - 1} - N$$
$$= \sqrt{N^2} - N = 0.$$

(b) If you follow through the steps given, you get

$$2(3X + 10) - 8, \quad \text{or} \quad 6X + 12.$$

Then division by 6 would give $X + 2$, and subtracting X would leave 2. So the sequence could finish 'divide by 6; take away the number you first thought of, and your answer is 2'. There are other possibilities, but this is one of the simplest.

(c) There are many possibilities. Here is one: Think of a number; quadruple it; add 12; halve the result; halve the result again; take away the number you first thought of, and your answer is 3.

Activity 37

You should try to have really good notes on this section as it is important later in the course.

Activity 38

Ensure that you include: the concepts of input and output; independent and dependent variables; and the use of X and Y to represent variables. You may also want to make some notes about the uses of various calculator keys.

Activity 39

In the formula $k = c + 273$, the doing operation is 'add 273'.

The corresponding undoing operation is 'subtract 273', and the undoing formula is

$$c = k - 273.$$

Activity 40

To make P the subject of the formula $J = 0.45P$, you need to undo the operation 'multiply by 0.45'. So divide both sides by 0.45 to get

$$\frac{J}{0.45} = \frac{0.45P}{0.45},$$

which simplifies to

$$P = \frac{J}{0.45}, \quad \text{or} \quad P = J \div 0.45.$$

Activity 41

(a) Kim should continue '... I must add 25 to my age'.

(b) To make x the subject, undo 'subtract 25'. So add 25 to both sides of $y = x - 25$:

$$y + 25 = x - 25 + 25.$$

Then

$$y + 25 = x, \quad \text{or} \quad x = y + 25.$$

Activity 42

(a) To make t the subject, divide both sides of $d = st$ by s and get

$$\frac{d}{s} = t, \quad \text{or} \quad t = \frac{d}{s}.$$

(b) To make d the subject, multiply both sides of $s = \dfrac{d}{t}$ by t and get

$$st = d, \quad \text{or} \quad d = st.$$

Activity 43

In

$$c = (f - 32) \div 18,$$

the *doing* steps are 'subtract 32' and 'divide by 1.8'.

The corresponding *undoing* steps are 'multiply by 1.8' and 'add 32'.

Therefore, to make f the subject, first multiply both sides by 1.8:

$$1.8c = f - 32.$$

Then add 32 to both sides:

$$1.8c + 32 = f.$$

Hence

$$f = 1.8c + 32.$$

Activity 44

(a) In

$$y = 2x + 1,$$

the doing steps are '×2' and '+1'.

The corresponding undoing steps are '−1' and '÷2'.

Therefore, to make x the subject, first subtract 1 from both sides:

$$y - 1 = 2x.$$

Then divide both sides by 2:

$$\frac{y - 1}{2} = x.$$

Hence

$$x = \frac{y - 1}{2}, \quad \text{or} \quad \frac{1}{2}(y - 1), \quad \text{or} \quad \frac{(y - 1)}{2}.$$

(b) In

$$y = \tfrac{1}{2}x - 1,$$

the doing steps are '÷2' and '−1'.

The corresponding undoing steps are '+1' and '×2'.

Therefore, add 1 to both sides:

$$y + 1 = \tfrac{1}{2}x.$$

Then multiply both sides by 2:

$$2y + 2 = x.$$

Hence

$$x = 2y + 2, \quad \text{or} \quad x = 2(y + 1).$$

(c) In

$$y = 2(x + 1),$$

the doing steps are '+1' and '×2'.

The corresponding undoing steps are '÷2' and '−1'.

Therefore, divide both sides of by 2:

$$\frac{y}{2} = x + 1.$$

Then subtract 1 from both sides:

$$\frac{y}{2} - 1 = x.$$

Hence

$$x = \frac{y}{2} - 1.$$

(d) In

$$y = \frac{x}{5},$$

the doing step is '÷5'.

The corresponding undoing step is '×5'.

Therefore, multiply both sides by 5:

$$5y = x.$$

Hence

$$x = 5y.$$

(e) In

$$y = \frac{x}{5} + 1,$$

the doing steps are '÷5' and '+1'.

The corresponding undoing steps are '−1' and '×5'.

Therefore, subtract 1 from both sides:

$$y - 1 = \frac{x}{5}.$$

Then multiply both sides by 5:

$$5y - 5 = x.$$

Hence

$$x = 5y - 5, \quad \text{or} \quad x = 5(y - 1).$$

(f) In
$$y = \frac{5x - 4}{3},$$
the doing steps are '$\times 5$', '-4' and '$\div 3$'.

The corresponding undoing steps are '$\times 3$', '$+4$' and '$\div 5$'.

Therefore, multiply both sides by 3:
$$3y = 5x - 4.$$
Add 4 to both sides:
$$3y + 4 = 5x.$$
Then divide both sides by 5:
$$\frac{3y + 4}{5} = x.$$
Hence
$$x = \frac{3y + 4}{5}.$$

(g) In
$$y = \frac{3(x + 1)}{4},$$
the doing steps are '$+1$', '$\times 3$' and '$\div 4$'.

The corresponding undoing steps are '$\times 4$', '$\div 3$' and '1'.

Therefore, multiply both sides by 4:
$$4y = 3(x + 1).$$
Divide both sides by 3:
$$\frac{4y}{3} = x + 1.$$
Then subtract 1 from both sides:
$$\frac{4}{3}y - 1 = x.$$
Hence
$$x = \frac{4}{3}y - 1.$$

(h) In
$$y = \frac{2}{3}(x + 2),$$
the doing steps are '$+2$', '$\times 2$' and '$\div 3$'.

The corresponding undoing steps are '$\times 3$', '$\div 2$' and '-2'.

Therefore, multiply both sides by 3:
$$3y = 2(x + 2).$$
Divide both sides by 2:
$$\frac{3}{2}y = x + 2.$$
Then subtract 2 from both sides:
$$\frac{3}{2}y - 2 = x.$$
Hence
$$x = \frac{3}{2}y - 2.$$

(i) In
$$y = \sqrt{x},$$
the doing step is 'take the square root'.

The corresponding undoing step is 'square'.

Therefore, square both sides:
$$y^2 = x.$$
Hence
$$x = y^2$$

(j) In
$$y = \frac{21}{x},$$
the doing steps are 'take the reciprocal' and '$\times 21$'.

The corresponding undoing steps are '$\div 21$' and 'take the reciprocal'.

Therefore, divide both sides by 21:
$$\frac{y}{21} = \frac{1}{x}.$$
Then take reciprocals of both sides:
$$\frac{21}{y} = \frac{x}{1}.$$
Hence
$$x = \frac{21}{y}.$$

(k) In
$$y = 6 + \frac{x}{20},$$
the doing steps are '$\div 20$' and '$+6$'.

The corresponding undoing steps are '−6' and '×20'.

Therefore, subtract 6 from both sides:

$$y - 6 = \frac{x}{20}.$$

Then multiply both sides by 20:

$$20y - 120 = x.$$

Hence

$$x = 20y - 120, \quad \text{or} \quad x = 20(y - 6).$$

(l) In

$$y = 2x^3,$$

the doing steps are 'cube' and '×2'.

The corresponding undoing steps are '÷2' and 'cube root'.

Therefore, divide both sides by 2:

$$\frac{y}{2} = x^3.$$

Then take the cube root of both sides:

$$\sqrt[3]{\frac{y}{2}} = x.$$

Hence

$$x = \sqrt[3]{\frac{y}{2}}.$$

(m) In

$$y = \frac{21}{x + 1},$$

the doing steps are '+1', 'take the reciprocal' and '×21'.

The corresponding undoing steps are '÷21', 'take the reciprocal' and '−1'.

Therefore, divide both sides by 21:

$$\frac{y}{21} = \frac{1}{x + 1}.$$

Take reciprocals of both sides:

$$\frac{21}{y} = x + 1.$$

Then subtract 1 from both sides:

$$\frac{21}{y} - 1 = x.$$

Hence

$$x = \frac{21}{y} - 1.$$

Activity 45

(a) Let y be the cost of the hire in guineas, and let x be the number of miles ridden. Now y equals the fixed charge plus the mileage cost: the fixed charge is 5 guineas; the mileage cost for x miles is x shillings, which equals $\frac{x}{21}$ guineas. Hence

$$y = 5 + \frac{x}{21}.$$

(b) The formula obtained in part (a) now has to be rearranged to make x the subject.

In

$$y = 5 + \frac{x}{21},$$

the doing steps are '÷21' and '+5'.

The corresponding undoing steps are '−5' and '×21'.

Therefore, subtract 5 from both sides:

$$y - 5 = \frac{x}{21}.$$

Then multiply both sides by 21:

$$21(y - 5) = x.$$

Hence

$$x = 21(y - 5).$$

The cost of hire was apparently 12 guineas, so $y = 12$. Substituting this value of y into the above equation gives

$$x = 21 \times 7 = 147.$$

Therefore the suspect travelled 147 miles on the hired motorbike.

The distance from central London to the edge of Norfolk is just over 70 miles, making it possible for the suspect to have committed the crime.

Activity 46

(a) As 8 kilometres are approximately equal to 5 miles, it follows that

$$1\,\text{km} = \frac{5}{8}\,\text{mile}$$

and

$$x\,\text{km} = \frac{5}{8}x\,\text{mile}.$$

But the distance from my house is 12 miles, hence

$$12 = \frac{5}{8}x, \quad \text{or} \quad \frac{5}{8}x = 12.$$

(b) In the above equation, the doing steps are '×5' and '÷8'.

The corresponding undoing steps are '×8' and '÷5'. This gives

$$x = 12 \times 8 \div 5$$
$$= 19.2.$$

So the distance that Yim had to travel to my house was just over 19 km.

(Check: $\frac{5}{8} \times 19.2 = 12$.)

Activity 47

Write the equation to be solved, with the variable on the left-hand side:

$$\frac{x}{21} + 5 = 120.$$

The doing operations in this equation are '÷21' and '+5'.

The corresponding undoing operations are '−5' and '×21'.

Therefore, subtract 5 from both sides:

$$\frac{x}{21} = 120 - 5 = 115.$$

Then multiply both sides by 21:

$$x = 115 \times 21 = 2415.$$

Hence the solution is $x = 2415$.

(Check: $2415/21 + 5 = 115 + 5 = 120$.)

Note that, in the remaining solutions, the doing and undoing steps are not listed. However, there are explanations of the operations performed on both sides of equations.

Activity 48

If you have checked your solutions by substituting them back into the original equations, you will know whether your answers are correct. They should be:

(a) $x = 2$ (subtract 1 and divide by 2);

(b) $x = 3$ (add 1 and multiply by 2);

(c) $x = \frac{3}{7}$ (add 3 and divide by 7).

Activity 49

The odd one out is $3x + 7 = 1$. All the other equations have a solution $x = 2$ and are rearrangements of each other.

Activity 50

The relevant equation is

$$^-20 = (f - 32) \div 1.8,$$

or

$$(f - 32) \div 1.8 = {}^-20.$$

To solve this equation, first multiply both sides by 1.8:

$$f - 32 = {}^-36.$$

Then add 32 to both sides:

$$f = {}^-4.$$

Hence the Fahrenheit equivalent of $^-20°C$ is $^-4°F$.

Activity 51

(a) In order to get all the x terms on the left-hand side of $3x = x + 24$, subtract x from both sides:

$$3x - x = x + 24 - x,$$

so
$$2x = 24.$$

Then divide both sides by 2, to obtain
$$x = 12.$$

Thus the solution is $x = 12$.

(b) To get all the x terms on the left-hand side of $5x - 6 = 2x$, subtract $2x$ from both sides:
$$5x - 2x - 6 = 2x - 2x$$
$$3x - 6 = 0.$$

Then add 6 to both sides, to obtain
$$3x - 6 + 6 = 0 + 6$$
$$3x = 6$$
$$x = 2.$$

Thus the solution is $x = 2$.

(c) To get all the x terms on the left-hand side of $4 + 9x = 18 + 2x$, subtract $2x$ from both sides:
$$4 + 9x - 2x = 18 + 2x - 2x$$
$$4 + 7x = 18.$$

Then subtract 4 from both sides, to obtain
$$4 + 7x - 4 = 18 - 4$$
$$7x = 14$$
$$x = 2.$$

Thus the solution is $x = 2$.

Activity 52

(a) When $x = {}^-40$, the expression on the right-hand side of the equation becomes $(x - 32) \div 1.8 = ({}^-72) \div 1.8 = {}^-40$.

This equals the left-hand side, as $x = {}^-40$.

So $x = {}^-40$ is a solution of the equation.

(b) When $x = 40$, the expression on the right-hand side of the equation becomes $(40 - 32) \div 1.8 = 8 \div 1.8 = \frac{40}{9}$.

This does not equal the left-hand side where $x = 40$.

So $x = 40$ is *not* a solution of this equation.

Activity 53

(a) To solve
$$3x = x + 4,$$

first subtract x from both sides:
$$2x = 4.$$

Then divide both sides by 2.

Hence
$$x = 2.$$

(b) To solve
$$x + 1 = 2x - 1,$$

first subtract x from both sides:
$$1 = x - 1.$$

Then add 1 to both sides:
$$2 = x.$$

Hence
$$x = 2.$$

(c) The equation $3(x + 1) = 2x - 1$ can be rewritten as
$$3x + 3 = 2x - 1.$$

To solve this equation, first subtract $2x$ from both sides:
$$x + 3 = -1.$$

Then subtract 3 from both sides.

Hence
$$x = {}^-4.$$

(d) To solve
$$5x \div 4 = x + 3,$$

first multiply both sides by 4:
$$5x = 4x + 12.$$

Then subtract $4x$ from both sides.

Hence
$$x = 12.$$

(e) To solve

$$7y + 1 = 17 - y,$$

first add y to both sides:

$$8y + 1 = 17.$$

Subtract 1 from both sides:

$$8y = 16.$$

Then divide both sides by 8.

Hence

$$y = 2.$$

(f) To solve

$$y \div 2 + 3 = y - 1,$$

first multiply both sides by 2:

$$y + 6 = 2y - 2.$$

Subtract y from both sides:

$$6 = y - 2.$$

Then add 2 to both sides.

Hence

$$8 = y, \quad \text{or} \quad y = 8.$$

When these solutions are checked, they all satisfy the original equations.

Activity 54

(a) Let Janet's present age be x years. Leroy's age is, therefore, $x - 10$. In two years' time, Janet will be $x + 2$ years old, and Leroy will be $x - 8$ years old. But Janet will then be twice as old as Leroy, so

$$x + 2 = 2(x - 8).$$

This equation can then be solved to find x, Janet's present age.

If $x = 18$, then the left-hand side of this equation is $x + 2 = 20$, and the right-hand side is $2(x - 8) = 2 \times 10 = 20$ also, confirming that Janet is now 18 years old.

(b) Suppose there are x people in my house; there must also be x pets. Therefore there are $2x$ heads, x tails and $2x + 4x$ feet. Now multiplying the number of heads by the number of tails gives the number of feet, so

$$2x \times x = 2x + 4x,$$

or

$$2x^2 = 6x.$$

If $x = 3$, then $2x^2 = 2 \times 9 = 18$, while $6x = 6 \times 3 = 18$. So $x = 3$ is a solution, indicating that I have three pets.

You may have spotted that $x = 0$ is also a solution of $2x^2 = 6x$, but at least one person lives in my house, namely me, so in this case it is not a valid solution.

Activity 55

(a) At a time t hours after 10 a.m., Alastair will be a distance of $75 + 50t$ miles from Cardiff. Bethan will be $400 - 60t$ miles from Cardiff. For Alastair and Bethan to meet, these distances must coincide, so the required equation must be

$$75 + 50t = 400 - 60t.$$

(b) To solve

$$75 + 50t = 400 - 60t,$$

add $60t$ to both sides and subtract 75. So

$$75 + 110t = 400,$$

which gives

$$110t = 325.$$

Hence

$$t = 325/110$$
$$= 2.91 \text{ (to 2 d.p.)}.$$

This gives a time of 2 hours 57 mins. So Alastair and Bethan should meet about 3 hours after 10 a.m., that is, at 1.00 p.m. (a good time for lunch!).

Activity 56

If the temperature x is to have the same numerical value on both scales, $f = x$ and $k = x$. So the equation to solve is

$$x = 1.8(x - 273) + 32,$$

or

$$x = 1.8x - 1.8 \times 273 + 32.$$

There are several ways of solving this equation. Here is one way.

To get all the x terms on the left-hand side, subtract $1.8x$ from both sides:

$$x - 1.8x = {}^-1.8 \times 273 + 32$$
$${}^-0.8x = {}^-1.8 \times 273 + 32.$$

Then divide both sides by $^-0.8$, to get

$$x = \frac{{}^-1.8 \times 273 + 32}{{}^-0.8} = 574.25.$$

So a temperature of $574.25°$F is also 574.25 on the Kelvin scale.

Activity 57

Let the man's life expectancy be x years. Then he spends $\frac{1}{15}x$ years as an infant, $\frac{2}{15}x$ years as a schoolboy, $\frac{1}{25}x$ years as a lover, $\frac{1}{5}x$ years as a soldier, $\frac{2}{5}x$ years as a justice, 10 years as a pantaloon, and 2 years in second childishness. The total length of his life is obtained by summing these up; together they must total x years. Thus

$$\tfrac{1}{15}x + \tfrac{2}{15}x + \tfrac{1}{25}x + \tfrac{1}{5}x + \tfrac{2}{5}x + 10 + 2 = x.$$

There are several ways to solve this. Here is one.

To get rid of the fractions, multiply everything by $5 \times 3 \times 5$:

$$5x + 10x + 3x + 15x + 30x + 750 + 150 = 75x.$$

Collecting the like terms on the left-hand side gives

$$63x + 900 = 75x.$$

To get all the x terms together on the right-hand side, subtract $63x$ from both sides:

$$900 = 12x.$$

Then divide both sides by 12:

$$900 \div 12 = x.$$

So

$$75 = x.$$

Therefore the life expectancy of a man is 75 years.

Check: The durations of the seven ages are

infant:	$\frac{1}{15} \times 75 = 5$ years
schoolboy:	$\frac{2}{15} \times 75 = 10$ years
lover:	$\frac{1}{25} \times 75 = 3$ years
soldier:	$\frac{1}{5} \times 75 = 15$ years
justice:	$\frac{2}{5} \times 75 = 30$ years
pantaloon:	10 years
second childishness:	2 years.

Then

$$5 + 10 + 3 + 15 + 30 + 10 + 2 = 75,$$

which tallies with a man's life expectancy of 75 years.

Activity 58

Your answer will be personal to you and your progress. Hopefully you will feel that you have achieved quite a lot even if you have not yet managed to grasp everything.

Activity 59

This activity is discussed on the associated audio band.

The solutions are

$$x \simeq 0.38 \quad \text{and} \quad x \simeq 2.62 \text{ (to 2 d.p.).}$$

Activity 60

(a) Someone who is touring will probably want to convert fairly large distances, perhaps several hundred miles, but will be unlikely to require very high accuracy. The table difference should be set to 10. The graph should cover the range from at least $M = 100$ to $M = 500$ or so.

(b) Distances on signposts for footpaths and cycle tracks are usually given to the nearest quarter of a mile; such distances are rarely more than 10 miles. The table difference should therefore be set to 0.25 or less. The graph should cover the range from $M = 0$ to $M = 12$, or so.

Activity 61

The graphs you obtain should look like this:

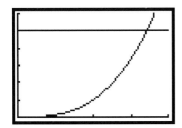

The solution, giving the cube root of 5, is 1.71 (to 2 d.p.).

[Check: $1.71^3 = 5.00$ (to 2 d.p.).
The value of $5^{1/3}$ (entered on the calculator as $5^{(1\div3)}$) is 1.709975947 (to 9 d.p.).]

Activity 62

(a) The patient weighs w kg. Dividing by 70 gives

$$\frac{w}{70}.$$

Taking the cube root gives

$$\sqrt[3]{\frac{w}{70}}.$$

Squaring the result gives

$$\left(\sqrt[3]{\frac{w}{70}}\right)^2.$$

Multiplying by 100 gives

$$d = 100\left(\sqrt[3]{\frac{w}{70}}\right)^2.$$

This is the formula for the daily dose d (in milligrams) of phenobarbital for a child weighing w kg.

(b) The graph below covers $w = 0$ to $w = 70$. (However, heavy patients, and babies, probably have to be treated differently, so care needs to be taken about using values at either end.)

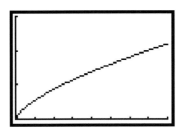

From the graph, when $w = 35$, d is about 63.

So the estimated daily dose of phenobarbital for a child weighing 35 kg is 63 mg per day (to the nearest milligram).

(c) Based on the formula obtained in part (a), the amended formula is

$$d = 200\left(\sqrt[3]{\frac{w}{70}}\right)^2,$$

where d is the four-hourly dose (in milligrams) of theophylline.

A table of doses of theophylline for children of different weights is shown below as a screendump.

X	Y₁	
10	54.655	
15	71.619	
20	86.76	
25	100.68	
30	113.69	
35	125.99	
40	137.72	

X=40

Activity 63

This may be an appropriate time to check that your Handbook sheet entries are complete so far, and also that you can still make sense of them. Make sure they are clearly referenced for ease of use—you may wish to include information such as the relevant section or page number in the unit. You may find that your ideas change or develop as you progress through the course. In that case, you may want to add additional notes to explanations and techniques or include different examples. Hopefully you will feel that you have progressed in algebra. However, you may feel that you want more practice. If so, look at the *Unit 8* questions in *Resource Book B*.

Acknowledgements

Grateful acknowledgement is made to the following sources for permission to reproduce material in this unit:

Cover

Train: Camera Press; map: reproduced from the 1995 Ordnance Survey 1:25 000 Outdoor Leisure Map with the permission of the Controller of Her Majesty's Stationery Office © Crown Copyright; other photographs: Mike Levers, Photographic Department, The Open University.

Index